Calcium Regulation by Calcium Antagonists

Calcium Regulation by Calcium Antagonists

Ralf G. Rahwan, EDITOR
Donald T. Witiak, EDITOR
The Ohio State University

Based on a symposium sponsored by the ACS Division of Medicinal Chemistry at the 182nd Meeting of the American Chemical Society, New York, New York, August 23–28, 1981.

ACS SYMPOSIUM SERIES 201

AMERICAN CHEMICAL SOCIETY
WASHINGTON, D. C. 1982

Library of Congress Cataloging in Publication Data

Calcium regulation by calcium antagonists.
(ACS symposium series, ISSN 0097–6156; 201)

"Based on a symposium sponsored by the ACS Division of Medicinal Chemistry at the 182nd Meeting of the American Chemical Society, New York, New York, August 23–28, 1981."

Includes bibliographies and index.

1. Calcium—Physiological effect—Congresses. 2. Calcium—Antagonists—Congresses.
I. Rahwan, Ralf G., 1941– . II. Witiak, Donald T. III. American Chemical Society. Division of Medicinal Chemistry. IV. American Chemical Society. National Meeting (182nd: 1981: New York, N.Y.) V. Series. [DNLM: 1. Calcium—Metabolism—Congresses. 2. Calcium—Antagonists and inhibitors—Congresses. 3. Ion channels—Metabolism—Congresses. 4. Calcium antagonists, Exogenous—Metabolism—Congresses. QV 276 C1443]

QP535.C2C265 1982 612'.01524 82–16451
ISBN 0–8412–0744–5 ACSMC8 201 1–208 1982

PRINTED IN THE UNITED STATES OF AMERICA

ACS Symposium Series

M. Joan Comstock, *Series Editor*

FOREWORD

The ACS SYMPOSIUM SERIES was founded in 1974 to provide a medium for publishing symposia quickly in book form. The format of the Series parallels that of the continuing ADVANCES IN CHEMISTRY SERIES except that in order to save time the papers are not typeset but are reproduced as they are submitted by the authors in camera-ready form. Papers are reviewed under the supervision of the Editors with the assistance of the Series Advisory Board and are selected to maintain the integrity of the symposia; however, verbatim reproductions of previously published papers are not accepted. Both reviews and reports of research are acceptable since symposia may embrace both types of presentation.

CONTENTS

PREFACE

ALTHOUGH THIS BOOK WAS CONCEIVED as a result of a Symposium entitled "Calcium Regulation and Drug Design," the topics presented at the Symposium have undergone revision, expansion, and updating, and additional chapters from invited authors have been incorporated into the present volume. This book, therefore, represents a collage of various topics involving the roles of calcium and calcium antagonists in health and disease. Unlike other excellent recent books on calcium antagonists, the present volume does not focus exclusively on the cardiovascular system, but covers other systems as well. The emphasis is on basic science, although clinical correlates are included wherever applicable. Numerous authorities in the field have contributed their expertise in their respective areas to the compilation of this book.

The opening chapter by Dr. Nayler attempts to create order from chaos by presenting a rational classification of membrane calcium channel blockers based upon their pharmacological effects on the cardiovascular system. Dr. Triggle then provides an overview of structure–activity studies with the calcium channel blockers, and emphasizes the chemical and pharmacological heterogeneity of this class of compounds. The latter fact provides a strong impetus for the adoption of the pharmacological classification of calcium channel blockers proposed by Dr. Nayler. A comprehensive review of the cardiovascular electrophysiological and hemodynamic effects of the calcium channel blockers is presented by Dr. Muir. Dr. Daniel and his colleagues provide insight into the role of calcium in the regulation of vascular and nonvascular smooth muscle contractility, and outline the methodological approaches for studying calcium fluxes, sequestration, and mobilization in smooth muscle.

While the first four chapters focus on the membrane calcium channels, the next two explore the intracellular compartment. Dr. Brostrom and his colleagues review present-day knowledge about the intracellular calcium receptor, calmodulin, while the Editors contribute an update on the basic and applied pharmacology of a novel class of intracellular calcium antagonists, the methylenedioxyindenes.

The roles of calcium and calcium antagonists in the central nervous system are dealt with in the next two chapters. From Dr. Leong Way's laboratory comes an overview of the mechanism of interaction between calcium and opioid alkaloids and peptides, while Dr. Ferrendelli reviews the evidence for an inhibitory effect of certain antiepileptic drugs on neuronal calcium conductance. Dr. Le Breton and his colleagues review the relationship of cyclic nucleotides and calcium in platelet function, while novel information on the salutary effects of the calcium antagonist nifedipine in atherosclerosis is presented from Dr. Henry's laboratory. Dr. Borowitz and coworkers review the role of calcium and the modulating influence of calcium antagonists on secretory systems.

We wish to acknowledge with gratitude the cooperation of the participants in the Symposium and the contributors to this volume with whom we had the distinct privilege of collaborating.

RALF G. RAHWAN
DONALD T. WITIAK
The Ohio State University
College of Pharmacy
Columbus, OH

June 1982

Calcium Antagonists: Classification and Properties

WINIFRED G. NAYLER

University of Melbourne, Department of Medicine, Austin Hospital, Heidelberg, Victoria, Australia

Calcium antagonists (slow channel blockers, slow Ca^{2+} antagonists) are a heterogeneous group of substances with widely differing tissue specificities, potency and properties. Some of them exhibit other properties in addition to that of Ca^{2+} antagonism. In general these drugs can be considered as a subgroup of a much larger group of compounds which impede the entry of Ca^{2+} irrespective of the route of entry. The 'slow channel blockers' can be subdivided on the basis of their differing tissue specificities. The purpose of this article is to explore the possibility that a classification which is based on differing tissue specificities may reflect differing modes of action.

Verapamil, nifedipine and diltiazem (Figure 1) belong to a relatively newly recognized group of drugs known collectively as "calcium antagonists", (1) "slow channel inhibitors", (2) or "calcium entry blockers" (3,4). Other substances which are now thought to belong to this group include niludipine, nimodipine, prenylamine, fendiline, caroverine, cinnarizine and perhexeline. The possibility of using these and other closely related substances in the management of a variety of cardiovascular disorders - including infarction, (5,6,7) arrhythmias, (8,9) angina, (10,11) hypertension (12) and hypertrophic obstructive cardiomyopathies (13) is now being considered. However the "calcium antagonists" that are currently available for such use differ from one another not only in terms of their chemistry, bio-availability and stability, (1,14) but also in potency, (1,14) tissue specificity and possibly in their precise mode of action (15,16,17). Because no attempt has yet been made to subclassify these drugs, the purpose of this article is to explore the problem of providing a suitable classification. To do this it is necessary to briefly discuss the physiological significance of the

0097-6156/82/0201-0001$06.00/0

Compound	Structure
Verapamil (mol.wt. 454.59)	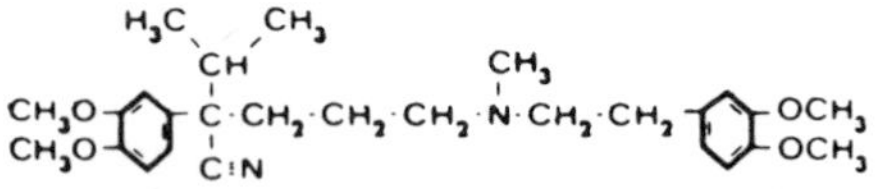 H_3C, CH_3, CH; CH_3O, CH_3O; C·CH_2·CH_2·CH_2·N(CH_3)·CH_2·CH_2; OCH_3, OCH_3; C⋮N
Compound **D-600** (gallopamil, mol.wt. 485.59)	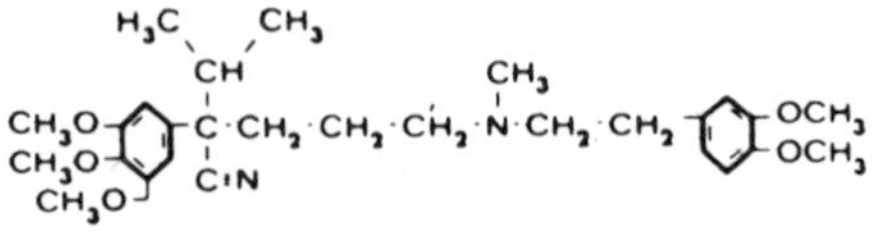 H_3C, CH_3, CH; CH_3O, CH_3O, CH_3O; C·CH_2·CH_2·CH_2·N(CH_3)·CH_2·CH_2; OCH_3, OCH_3; C⋮N
Nifedipine (mol.wt., 346.34)	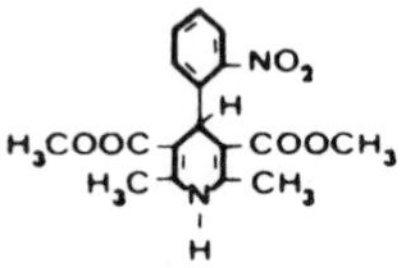 NO_2; H; H_3COOC, $COOCH_3$; H_3C, CH_3; N; H
Niludipine (mol.wt. 490.55)	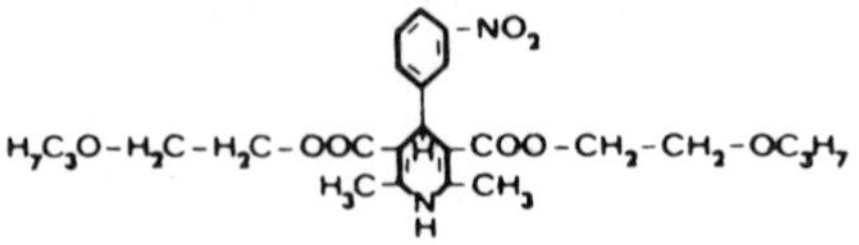 NO_2; $H_7C_3O-H_2C-H_2C-OOC$, $COO-CH_2-CH_2-OC_3H_7$; H_3C, CH_3; N; H
Nimodipine (mol.wt. 418.45)	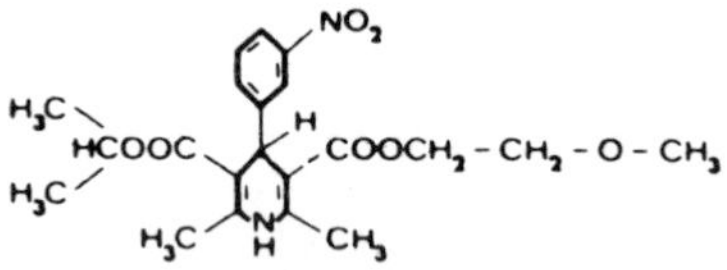 NO_2; H_3C, HCOOC, H_3C; H; $COOCH_2-CH_2-O-CH_3$; H_3C, CH_3; N; H
Diltiazem (mol.wt. 414.52)	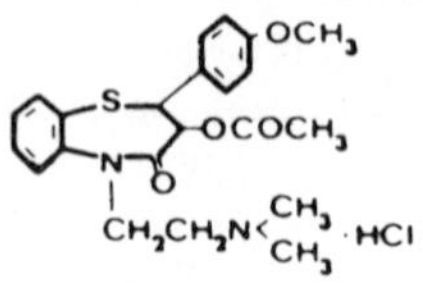 OCH_3; S; $OCOCH_3$; N; O; $CH_2CH_2N<(CH_3)(CH_3)$ ·HCl

Figure 1. Structural formulas of some Ca^{2+} antagonists and their derivatives.

"slow Ca^{2+} current" (18) and the associated ion-selective channels.

The Slow Ca^{2+} Current

The slow Ca^{2+} current in heart muscle.

In normal heart muscle excitation involves the activation of two distinct inward currents (19,20). The first of these currents is carried by Na^+ and is recorded as the fast upstroke of the action potential. The Na ions move across the cell membrane through voltage-activated "channels" that are highly, but not totally, selective for Na^+. The second inward current is carried mainly (21,22), but not exclusively (23), by Ca^{2+}. It is activated slowly, contributes to the plateau phase of the action potential and is known as "the slow Ca^{2+} current". The Ca ions involved pass across the cell membrane through channels that are highly selective for Ca^{2+}. Like their Na^+ counterparts, the Ca^{2+}-selective channels are voltage activated but their threshold of activation (about-55mV) is higher than that of the Na^+ channels (-35mV). It is possible to envisage these "channels" as being pore-like structures in the plasmalemma, each pore having its own set of "activation" and "inactivation" gates. In this analogy voltage activation can be likened to the opening of "gates" which are closed during the resting state (24). Taking this hypothesis one step further it can be argued that the normal opening and closing of these "gates" involves voltage-dependent changes in the configurational state of the membrane. This same argument applies irrespective of whether the ion-selective "channels" are "pore-like" structures or a particular combination or organization of the membrane proteolipids that facilitates the inward movement of certain ions. Within this framework drugs which interact with the cell membrane may alter the configurational state of the membrane, thereby altering the ion selectivity or responsiveness of the "gated" channels and proteolipid complexes.

The slow Ca^{2+} current and excitation-contraction coupling.

The slow Ca^{2+} current accounts for the entry into the cytosol of between 5 and 10 μmoles Ca^{2+}/kg heart weight/beat (25). This is approximately one tenth of the Ca^{2+} needed to activate contraction (26). Probably the Ca ions that enter as the main charge carriers for the slow current serve as a trigger for the mobilization of Ca^{2+} from the intracellular stores (27,28). However, since the magnitude of the mechanical response varies according to the extracellular concentration of Ca^{2+} (29) it is probably the current-carrying Ca ions of the slow inward current which determine the amount of Ca^{2+} mobilized for interaction with the myofilaments. In other words, the Ca^{2+}-induced "trigger" release of Ca^{2+} from the intracellular storage sites simply provides an internal amplification factor. Skeletal muscle differs from cardiac muscle in that it mobilizes all the Ca^{2+} it requires for excitation-contraction coupling directly from its own intracellular stores.

The slow Ca^{2+} current in smooth muscle, nodal and conducting tissues.

The occurrence of "Ca^{2+}-selective", voltage activated "channels" is not limited to the myocardium. They occur in most smooth muscle cells - including those in the coronary, cerebral and peripheral vasculature (30). The normal activity of pacemaker, nodal and conducting tissues (31) is also largely dependent upon them. Because of their widespread distribution it follows that substances which affect the functioning of these channels will have a profound effect on the circulation. By contrast, skeletal muscle is relatively unaffected (32).

Specificity of slow channels for Ca^{2+}.

Although Ca^{2+} is the main current carrying ion for the slow inward current part of the current is carried by Na^+ (23). Certain divalent cations, including Ba^{2+} and Sr^{2+}, can substitute for Ca^{2+} as the main charge carrier (33). Others, including Ni^{2+}, Co^{2+}, and Mn^{2+} are inhibitory (34).

The Identification of Ca^{2+}-antagonists

Techniques which are currently being used to identify substances that alter slow channel transport involve measurements of the height and duration of the action potential, monitoring the rate of uptake of radioactively labelled Ca^{2+}, and electrophysiological techniques that involve suppression of the fast Na^+ current. Use of these techniques is based on the tacit assumption that the drugs we are dealing with act only on the slow channels. Later in this chapter we will summarize the data which suggests that such an assumption may no longer be justified. For the moment, however, it is pertinent to work within this framework.

(a) Height and duration of the action potential.

To illustrate the use of changes in the configuration of the action potential to establish the presence or absence of "Ca^{2+} antagonism" we will concentrate on the cardiac action potential. The currents which contribute to the height and duration of the cardiac action potential are complex (20). In addition to the *inward* currents already referred to there are at least two and possibly three *outward* (35) K^+ currents. Accordingly, unless a substance is *specific* for only the inward Ca^{2+} current studies which depend upon the monitoring of changes in the height and duration of the action potential are not particularly helpful in identifying either inhibitors or activators of slow channel transport. This difficulty is further compounded by the fact that the subsarcolemmal concentration of Ca^{2+} influences the outward K^+ currents, (35) and hence the shape and duration of the action potential.

(b) Isotopic Techniques

The use of radioactively labelled Ca^{2+} to monitor changes in the magnitude and duration of the slow Ca^{2+} current is also difficult, partly because of the small amounts of Ca^{2+} that are involved. When considered on a beat to beat basis the Ca^{2+} that enters heart muscle cells by way of the slow Ca^{2+} current represents less than 2 percent of the total tissue Ca^{2+} and if, as seems probable, this small component is rapidly recycled to the exterior its accurate detection during the time course of the action potential presents substantial technical difficulties. There is, however, another and more serious objection to the use of labelled Ca^{2+} for this purpose. This difficulty centres around the fact that Ca^{2+} can enter the myocardium and other excitable cells through several different routes - including in exchange for Na^+, by passive diffusion, and (Figure 2) in exchange for K^+. Consequently even if the uptake of radioactively labelled Ca^{2+} is reduced there can be no certainty that the particular route of entry that is being affected is "the slow channels".

(c) Electrophysiological techniques

These involve the use of techniques to suppress the fast inward Na^+ current. Either a voltage can be applied so that the transmembrane potential difference is clamped above the level at which the Na^+ current is activated, (36) or tetrodotoxin (37) - which specifically inhibits the Na^+ current, can be added, or the membrane can be depolarized by raising the external K^+ (38,39). Under each of these conditions the slow inward current can be activated by adding aminophylline or isoproterenol, (39) and by electrical stimulation.

Substances That Alter Slow Channel Transport

Theoretically, substances or interventions that alter slow channel transport may do so in a variety of ways:-

(a) they may alter the amount of Ca^{2+} which is available to act as the charge carrier;

(b) they may, by interacting with the cell surface, evoke a configurational change which either facilitates or impedes the approach of Ca^{2+} to the channels; alternatively -

(c) the configurational change in the membrane may induce a

change in the Ca^{2+}-carrying capacity of each channel; or -

(d) the number of channels that are operative at any given time may be affected. This could be achieved by altering the threshold of activation, changing the kinetics of channel activation and/or recovery, by facilitating the formation of "de novo" channels or by activating "sleeping" channels.

Activators of slow channel transport.

The catecholamines, including norepinephrine, epinephrine and isoproterenol augment the slow Ca^{2+} current (39,40). They do this by increasing the number of channels that are activated at a given voltage, without affecting the rate of channel activation or deactivation (40). Cyclic AMP has the same effect (41). This ability of the catecholamines to recruit new channels may indicate the existence of a heterogenous population of voltage-activated channels, one population being under cyclic AMP control. Alternatively there may be several different states of activation for each channel. Irrespective of which, if either, of these alternatives is correct, it follows that the amount of slow channel activity that is available at any one moment is influenced by the circulating level of catecholamine. The recent discovery of 'calciductin', a protein that appears to be associated with the slow channels and which can be phosphorylated by a cyclic AMP-dependent pathway points towards the possibility of heterogeneity within the channels. There are other reasons for believing that the channels themselves must be heterogenous. Thus, for example, the drugs we are discussing have no inhibitory effect on the Ca^{2+}-dependent excitation-induced release of norepinephrine from the sympathetic nerve terminals.

Inhibitors of slow channel transport.

Many substances inhibit slow channel transport. In addition to the divalent cations already cited (Mn^{2+}, Co^{2+} and Ni^{2+}), protons, La^{3+}, the metabolic inhibitors cyanide and dinitrophenol, are effective inhibitors (24). Other inhibitory agents include acetylcholine, (42) papaverine, (43) pentobarbital, lidoflazine, (44) and adenosine (45) as well as verapamil, nifedipine and diltiazem (1). Precisely how many of these substances interfere with slow channel transport is unknown, although in the case of the metabolic inhibitors we can probably account for their effect in terms of the energy requirements (24) needed for maintaining the configurational state of the cell membrane compatible with the maintenance of normal slow channel ultrastructure.

Classification of Slow Channel Blockers (or Antagonists)

Fleckenstein (1) originally classed some of the substances listed in Table I as "calcium antagonists" on the basis of two requirements:

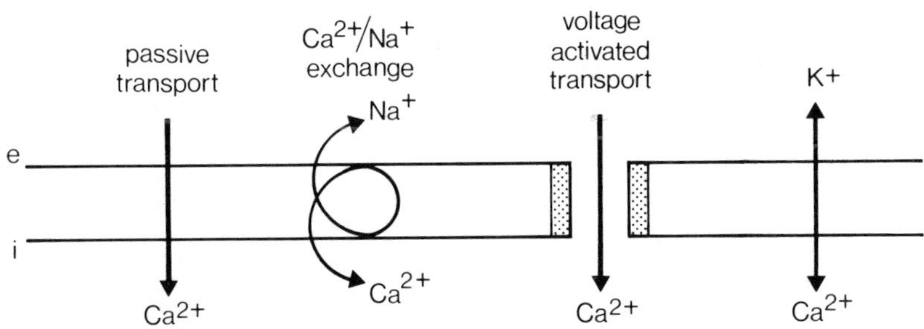

Figure 2. Schematic representation of possible routes of Ca^{2+} entry into a myocardial cell.

Table I: Substances Classed as Ca^{2+} Antagonists

	References
Verapamil	1
methoxyverapamil (D-600)	1
prenylamine	1
nifedipine	1
lidoflazine	44
nimodipine	52,53
diltiazem	1
bepridil	68
caroverine	70
niludipine	53
fendiline	1

(a) the predominant characteristic of these substances is their ability to inhibit the slow Ca^{2+} current; and
(b) this inhibition could be overcome by adding Ca^{2+}.

These criteria still apply. However, the continued unqualified use of the term "calcium antagonist" requires reappraisal because of its lack of specificity with respect to the site and precise mode of drug action. Thus "calcium antagonism" can be expressed at a variety of sites, including the cell membrane, the myofibrils, the sarcoplasmic reticulum and the mitochondria. When used in therapeutic concentrations, however, the drugs we are discussing express their "calcium antagonistic" properties at only one site - the cell membrane. Even at the cell membrane there are other ways in which substances can interfere with transmembrane Ca^{2+} movements - apart from the entry of Ca^{2+} through the voltage-activated "channels". Possibly, therefore, there is some merit in considering drugs of the type shown in Table I as being a subgroup of a much larger group of drugs which, for want of a better term, may be called "Ca^{2+}-entry blockers" or "Ca^{2+}-entry antagonists" (34). This group of drugs - "the Ca^{2+} entry blockers" or "Ca^{2+}-entry antagonists" (Figure 3) would include any drug which impedes the inward movement of Ca^{2+}, irrespective of the route of entry. The known routes of Ca^{2+} entry into cardiac and smooth muscle cells include by passive diffusion, in exchange for Na^{+}, in exchange for K^{+}, and through the voltage-activated, ion selective channels we have been discussing. As far as cardiac and smooth muscle cells are concerned, therefore it is possible that four different sub groups of Ca^{2+}-entry blockers (or antagonists) will ultimately become available. However, the drugs which are currently available are specific only for the subgroup that involves the influx of Ca^{2+} through the voltage activated, ion selective channels. Since these channels are slowly activated relative to the channels that selectively facilitate the rapid influx of Na^{+} during the fast upstroke phase of the action potential it may be more appropriate to refer to these substances as "inhibitors of slow channel transport". An alternative term - "Ca^{2+} channel blocker" has already appeared in the literature but may be inappropriate because the slow channels are not totally selective for Ca^{2+}. They also admit some Na^{+} and the resultant "slow" Na^{+}-dependent current is blocked by some of the currently available antagonists, including verapamil and methoxy verapamil (23). Presumably as new Ca^{2+} entry blocking drugs are developed substances that specifically inhibit the influx of Ca^{2+} through routes of entry other than the slow channels will become available. For example, substances that specifically inhibit the entry of Ca^{2+} in exchange for Na^{+} or K^{+} may be developed. Such substances could be of clinical importance because these other routes of Ca^{2+} entry may be involved (43) in the massive influx of Ca^{2+} that occurs (46) when flow is re-introduced to a previously ischaemic zone - as may occur, for example, when coronary vasospasm is relieved, a thrombus is dissolved or removed, or revascularization takes place.

Subclassification

Substances which inhibit slow channel transport can be conveniently subdivided into two major subgroups, on the basis of their chemistry. Thus there are the inorganic (Co^{2+}, Mn^{2+}, La^{3+}) and the organic inhibitors. The organic inhibitors can be further subdivided into three main classes, on the basis of their tissue specificity. According to this scheme drugs which predominately affect slow channel activity in the myocardium - example verapamil, can be classed as having strong Class I activity (47). Class II could include those drugs which are most effective in blocking slow channel transport in vascular smooth muscle. Nifedipine would provide the prototype for this subgroup (47). Class III would include those substances which are most potent in blocking slow channel transport in pacemaker, nodal and conducting tissues. Verapamil, therefore would have strong Class III activity, (8) whereas nifedipine would have only weak Class III activity (48,49). Diltiazem (11) exhibits strong Class II activity (50,51). These relative activities are summarized in Table II. Working from such a classification it is relatively easy to determine which particular slow channel blocker should be selected for use in a particular situation. For example, because of its strong Class III activity verapamil is the drug of choice if it is the Ca^{2+} currents in nodal, pacemaker or conducting tissues that are to be suppressed. On the other hand nifedipine may be the drug of choice for relieving coronary artery spasm - an activity which would reflect its strong Class II activity.

Class II drugs can be further subdivided into at least three subgroups (Figure 3). For example the effect of diltiazem on the slow channels is more marked in the smooth muscle cells of the coronary (51) than the peripheral vasculature. Nimodipine (52) acts preferentially on slow channel transport in the cerebral vessels, where it blocks thromboxane-induced contractions (53). By contrast, lidoflazine is more potent in the periphery. Even within the coronary circulation there is room for further subclassification according to whether it is the small or large vessels (Figure 3) that are being affected (45). For example, adenosine, which is a relatively weak Ca^{2+} entry blocker, acts on smooth muscle cells in the small coronary arteries, whereas diltiazem and nifedipine act preferentially on the large vessels.

Why the various organic inhibitors differ with respect to their preferred site of action is unknown. Is it because of their different chemical structure? Or are the slow Ca^{2+} channels themselves tissue specific? Some substances - eg. pentobarbitone and adenosine, which are relatively weak slow channel blockers do not display tissue specificity with respect to their Ca^{2+} antagonism. Indeed in these substances it is questionable

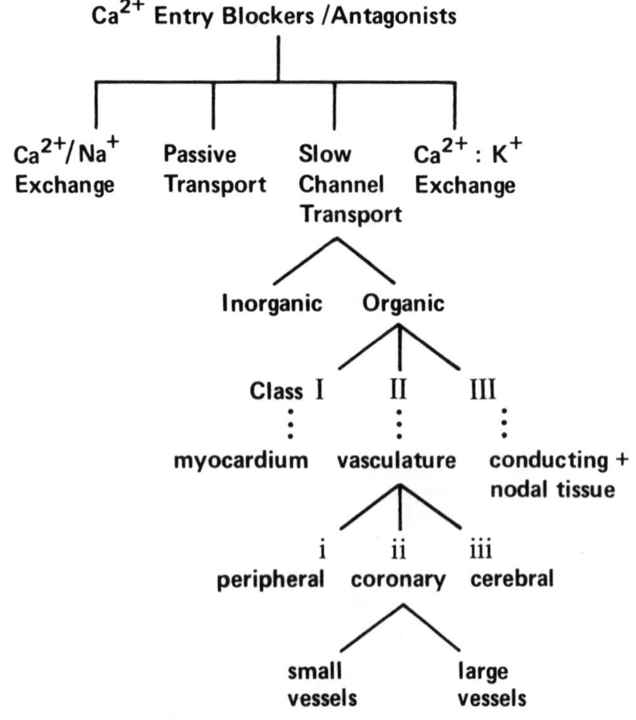

Figure 3. Proposed subdivision of Ca^{2+}-entry blocking drugs.

Table II: Subdivision of Slow Channel Inhibitors

Class	Slow Channel Inhibitor			
	Verapamil	Nifedipine	Diltiazem	Lidoflazine
I	strong	weak	weak	absent
II	weak	strong	strong	strong
III	strong	absent	weak	absent

whether, because of their non-specificity, these substances should be classified as slow channel blockers.

Site and Mode of Action.

In the classification shown in Figure 3 the organic and inorganic inhibitors of slow channel transport have been grouped together as subgroups of a common type of Ca^{2+} entry blocker. This does not mean that they have a common site of action. For instance, the inorganic inhibitors - Mn^{2+}, $Ni^{2+}$and Co^{2+} - act at the outer surface of the cell membrane, where they compete with Ca^{2+} for binding sites (34,54). By contrast the organic inhibitors - eg. verapamil, D-600 appear to act at the cytosolic surface (55,56). Even the organic inhibitors differ in their precise mode of action. For example, nifedipine (46) reduces the number of channels that are operational at a given time without altering the kinetics of channel activation or inactivation, whereas verapamil (17) alters the kinetics of channel reactivation so that recovery is delayed after a prior period of activity. Possibly this effect of verapamil is associated with its ability to displace Ca^{2+} from superficially located binding sites (47).

Theoretically it should be possible to classify the slow channel blockers according to whether or not they affect the kinetics of slow channel transport. In this way we could readily separate the nifedipine type of drugs from those that are more like verapamil. Alternatively can these drugs be subgrouped according to their chemistry? This possibility seems to be remote. Thus diltiazem is a benzothiazepine derivative (Figure 1), nifedipine is derived from dihydropyridine whilst verapamil has some structural features in common with papaverine.

Structure activity relationships are available for verapamil and nifedipine but not, as yet, for diltiazem.

The activity of verapamil as a slow channel inhibitor depends upon the presence of the two benzene rings and the tertiary amino nitrogen. Loss of the tertiary amino nitrogen - as occurs, for example, upon quarternization, results in a total loss of activity, whilst substitutions within the benzene rings result in a decreased potency. There are several ways of interpreting this data. The tertiary amino nitrogen may be acting as an essential "spacer unit", ensuring that the distance between the two aromatic rings is optimal for "receptor" occupancy. Alternatively permanent ionization of the nitrogen, as occurs upon quarternization, may result in the molecule being unable to penetrate into the membrane. Now if, as seems likely, the binding site for verapamil is located at or near the cytosolic surface of the membrane permanent ionization could prevent the molecule from reaching its binding site, thus rendering it inactive.

Like verapamil, nifedipine contains two aromatic rings (Figure 1), but this seems to be the only property shared by these two molecules. In the case of nifedipine substitutions at the ester position of the N-heterocyclic ring that increase its lipid solubility decrease its potency as a slow channel inhibitor. A low lipid solubility, therefore, must be of some importance. In addition, and although the presence of the NO_2 group at the ortho position is not essential, substitutions at the ortho, meta or para positions of the benzene ring result in a decreased activity, particularly if electron donating or electron withdrawing radicles are added, or if long side arms are introduced. It seems likely, therefore, that so far as nifedipine is concerned low lipid solubility and its steric configuration are both important.

The heterogeneity of the Ca^{2+} antagonists.

We have already mentioned the possibility of the Ca^{2+} channels being heterogeneous. There is increasing evidence that the heterogeneity of the slow channel blockers may extend beyond their chemistry to their precise mode of action. Thus whilst Fleckenstein's original recognition of these compounds was based on their ability to inhibit slow channel transport in the myocardium the potency of the vasodilator activity which some of these substances exert may involve more than slow channel inhibition. Indeed it is always pertinent to remember that the inhibitory effect of these substances on membrane activity in smooth muscle cells has always relied upon the presence of an abnormal extracellular ionic environment - usually a raised K^+. Data is beginning to accumulate now which indicates that some of these substances - including cinnarizine, flunarizine, fendiline (57), and felodipine (58) may interact with calmodulin to reduce its Ca^{2+} binding activity. Since the interaction of the calmodulin -Ca^{2+} complex with myosin light chain kinase plays an important role in determining the contractile state of smooth muscle cells, this intracellular action may account, in part at least, for the greater sensitivity which some of these drugs exhibit for smooth muscle cells - that is these highly potent vasodilators may interact with the Ca^{2+}-calmodulin complex as well as inhibiting slow channel transport. This is an intriguing possibility and it may provide the basis for the division of these substances as shown in Figure 3. Thus Class II compounds could be those which act predominately within the cell, whereas Class I compounds could include those which act predominately on the voltage activated slow channels. An intracellular site of action need not be limited to an effect on the Ca^{2+}-binding activity of calmodulin. Thus there is some evidence to support the possibility that diltiazem which, because of its powerful coronary vasodilator activity (59) would be included in the Class II subdivision of the slow channel blockers, inhibits the release of Ca^{2+} from intracellular storage sites in smooth muscle cells (60).

How then can we account for the Class III (Figure 3) compounds? To do this we may have to take into account yet another property of these compounds - that is the ability of some of them, including verapamil and diltiazem (59) to interfere with the fast Na^+ channels. This is an activity which is expressed at relatively high dose levels but which is absent from those compounds (nifedipine, niludipine and nimodipine) which exhibit strong Class II but no Class III (Figure 3) activity.

Other Properties of Slow Channel Blockers

Although suppression of the voltage-activated inward displacement of Ca^{2+} is undoubtedly the most widely discussed property of the substances listed in Table I, some of these drugs exhibit other properties that may be of clinical relevance. For example verapamil (61,62) and nifedipine bind to α receptors in brain, neuroblastomaglioma hybrid (63) and cardiac muscle cells (64). Verapamil and methoxyverapamil also bind fairly tightly to muscarinic receptors (65). Verapamil and D-600 have an inhibitory effect on Na^+ and K^+ conductance, (23,63) an effect which apparently is not shared by nifedipine. Diltiazem slows the release of Ca^{2+} from smooth muscle sarcoplasmic reticulum (66). As far as the circulation is concerned, however, these other properties of the slow channel blockers are probably insignificant when compared with their inhibitory effect on slow channel and possibly on calmodulin-Ca^{2+} binding activity.

The Identification of Ca^{2+}-Antagonist Binding Sites

Possibly, and in the very near future, we may be able to subdivide the Ca^{2+}-entry blockers in terms of their interaction with specific membranes-located binding sites. Thus, in a recent study (67) highly specific binding sites have been identified in rabbit cardiac homogenates for H^3-labelled nifedipine. Other Ca^{2+} entry blockers have been found to compete with ^3H-nifedipine for these binding sites, the order of potency being nifedipine >> D-600 = verapamil > cinnarizine. Interestingly, diltiazem and perhexeline were found not to inhibit ^3H nifedipine binding. Thus there is now good reason for believing that it is not only the slow Ca^{2+} channels themselves which are heterogeneous: the drugs which we now believe to interact with them are also heterogeneous not only in their chemistry, but also in their mode of action. As might have been expected substances with Ca^{2+} antagonist properties are now being isolated from some of the traditional herbal cures - one such is 'tanshinone', an effective antianginal compound that has been isolated from the roots of Salvia miltiorrhiza Bunge used as Dan Shen in traditional Chinese medicine (69).

Conclusion

The currently available slow channel inhibitors (or Ca^{2+} antagonists) can be conveniently classed as a subgroup of a large group of compounds which block the entry of Ca^{2+} into cells. Within this subgroup further subclassification is possible, based on tissue specificity. Whilst such a classification does not explain why these substances differ with respect to their tissue specificity, it provides a useful framework for comparing their different modes of action.

Literature Cited

1. Fleckenstein, A. Specific inhibitors and promoters of calcium action in the excitation-contraction coupling of heart muscle. In: Calcium and the Heart, eds. Harris, P and Opie, L. Academic Press, London 1970, 135-188.
2. Katz, A.M.; Reuter, H. Am. J. Cardiol. 1979, 44, 188-190.
3. Nayler, W.G. European Heart J. 1980, 1, 225-237.
4. Nayler, W.G.; Grinwald, P. Fed. Proc. 1981, 40, 2855-2861.
5. Christlieb, I.Y.; Clark, R.E.; Nora, J.D.; Henry, P.D.; Fischer, A.E.; Williamson, J.R.; Sobel, B.E. Am. J. Cardiol 1979, 44, 825-831.
6. Reimer, K.A.; Lowe, J.W.; Jennings, R.B. Circ. 1977, 55, 581-587.
7. Nayler, W.G.; Ferrari, R.; Williams, A. Am. J. Cardiol. 1980, 46, 242-248.
8. Krikler, D.M.; Spurrell, R.A.J. Postgrad. Med. J. 1974, 50, 447-453.
9. Zipes, D.P.; Fischer, J.C. Circ. Res. 1974, 34, 184-192.
10. Andreasen, F.; Boye, E.; Christoffersen, D. Europ. J. Cardiol. 1975, 2, 443-452.
11. Gunther, S.; Green, L.; Muller, J.E.; Mudge, G.H.; Grossmann, W. Am. J. Cardiol. 1979, 44, 793-797.
12. Lewis, G.R.J.; Morley, K.D.; Lewis, B.M.; Bones, P.J. New Zealand Med. J. 1978, 612, 351-354.
13. Rosing D.R.; Kent, K.M.; Maron, B.J.; Condit, J.; Epstein, S.E. Chest. 1980, 78, (Suppl.1) 239-247.
14. Fleckenstein, A. Ann. Rev. Pharmacol. Toxicol. 1977, 17, 149-166.
15. Raschack, M. Arzneim. Forsch. 1976, 26, 1330-1333.
16. Kohlhardt, M.; Fleckenstein, A. Naunyn. Schmiedeberg's Arch. Pharmacol. 1977, 298, 267-272.
17. Kohlhardt, M.; Mnich, Z. J. Mol. Cell. Cardiol. 1978, 10, 1037-1054.
18. Rougier, O.; Vassort, G.; Garnier, D.; Gargouil, Y.M.; Coraboeuf, E. Pfluegers Arch. 1969, 308, 91-110.
19. New, W.; Trautwein, W. Pfluegers Arch. 1972, 334, 1-23.
20. Coraboeuf, E. Am. J. Physiol. 1978, 234, 1101-1116.
21. Beeler, G.W. Jr.; Reuter, H. J. Physiol. (Lond) 1970, 207, 191-209.

22. Reuter, H. Ann. Rev. Physiol. 1979, 41, 413-424.
23. Kass, R.S.; Tsien, R.W. J. Gen. Physiol. 1975, 66, 169-192.
24. Sperelakis, N.; Schneider, J.A. Am. J. Cardiol. 1978, 37, 1079-1085.
25. Langer, G.W. J. Mol. Cell. Cardiol. 1980, 12, 231-240.
26. Solaro, R.J.; Wise, R.M.; Shiner, J.S.; Briggs, F.N. Circ. Res. 1974, 34, 525-530.
27. Nayler, W.G. J. Clin. Orthop. 1966, 46, 157-186.
28. Fabiato, A.; Fabiato, F. An. New York Acad. Sci. 1978, 307, 491-522.
29. Ringer, S. J. Physiol. 1883, 4, 29-42.
30. Godfraind, T.; Kaba, A. Arch. Int. Pharmacodyn. Ther. 1972, 196, 35-49.
31. Noble, D. Oxford Clarendon Press. 1975.
32. Shine, K.I.; Serena, S.D.; Langer, G.A. Am. J. Physiol. 1971, 221, 1408-1417.
33. Kohlhardt, M.; Haastert, H.P.; Krause, H. Pfluegers Arch. fur die gesamte Physiologic. 1973, 342, 125-136.
34. Kohlhardt, M.; Mnich, Z.; Haap, K. J. Mol. Cell. Cardiol. 1979, 12, 1227-1244.
35. Bassingthwaighte, J.B.; Fry, C.H.; McGuigan, J.A.S. J. Physiol. 1976, 262, 15-37.
36. Beeler, G.W.; Reuter, H. J. Physiol. 1970, 207, 165-190.
37. Shigenobu, K.; Sperelakis, N. J. Mol. Cell. Cardiol. 1971, 3, 271-286.
38. Pappano, A.J. Circ. Res. 1970, 27, 379-390.
39. Schneider, J.A.; Sperelakis, N. J. Mol. Cell. Cardiol. 1975, 7, 249-273.
40. Reuter, H.; Scholz, H. J. Physiol. 1977, 264, 49-62.
41. Watanabe, A.B.; Besch, H.R. Circ. Res. 1974, 35, 316-324.
42. Ikemoto, Y.; Goto, M. J. Mol. Cell. Cardiol. 1977, 9, 313-326.
43. Schneider, J.A.; Brooker, G.; Sperelakis, N. J. Mol. Cell. Cardiol. 1975, 7, 867-876.
44. Van Nueten, J.M.; Wellens, D. Arch. Int. Pharm. Ther. 1979, 242, 329-331.
45. Harder, D.R.; Belardinelli, L.; Sperelakis, N.; Rubio, R.; Berne, R.M. Circ. Res. 1979, 44, 177-182.
46. Shen, A.C.; Jennings, R.B. Am. J. Path. 1979, 67, 417-440.
47. Nayler, W.G.; Szeto, J. Cardiovasc. Res. 1972, 6, 120-128.
48. Ebner, F.; Donath, M. Mode of action and efficacy of nifedipine in 4th International Adalat Symposium. ed. P. Peuch, R. Krebs. Publ. Excerpta Medica. Amsterdam, 1980, 25-34.
49. Rowland, E.; Evans, T.; Krikler, D. Brit. Heart. J. 1979, 42, 124-127.
50. Kusukawa, R.; Kinoshita, M.; Shimoto, Y.; Tomonaga, G.; Hoshina, T. Arzneim. Forsch. 1977, 21, 878-883.
51. Henry, P.D.; Shuchleib, R.; Borda, L.R.; Roberts, R.; Williamson, J.R.; Sobel, B.E. Circ. Res. 1978, 43, 372-380.
52. Towart, R.; Kazda, S. Brit. J. Pharmacol. 1979, 67, 409P.

53. Towart, R.; Perzborn, E. Europ. J. Pharm. 1981, 69, 213-215.
54. Kohlhardt, M.; Wais, V. J. Mol. Cell. Cardiol. 1979, 11, 917-922.
55. Payet, M.D.; Schanne, O.F.; Ruiz-Ceretti, E.; Demers, J.M. J. Mol. Cell. Cardiol. 1980, 12, 187-200.
56. Payet, M.D.; Schanne, O.F.; Ruiz-Ceretti, E. J. Mol. Cell. Cardiol. 1980, 12, 635-638.
57. Spedding, M. Brit. J. Pharm. 1982, 75, 25P.
58. Bostrom, S.L.; Ljung, B.; Mardh, S.; Forsen, S.; Thulin, E. Nature. 1981, 292, 777-778.
59. Henry, P.D. Amer. J. Cardiol. 1980, 46, 1047-1058.
60. Takenaga, H.; Magaribuchi, T.; Nakajima, H. Japan J. Pharm. 1979, 28, 457-464.
61. Fairhurst, A.S.; Whittaker, M.L.; Ehlert, F.J. Biochem. Pharm. 1980, 29, 155-162.
62. Glossmann, H.; Hornung, R. Mol. Cell. Endrocrinol. 1980, 19, 243-251.
63. Atlas, D.; Adler, M. Proc. Natl. Acad. Sci. 1981, 78, 1237-1241.
64. Nayler, W.G.; Thompson, J.E.; Jarrott, B. J. Mol. Cell. Cardiol. In Press, 1982.
65. Cavey, D.; Vincent, J.P.; Lazdunski, M. FEBS letters. 1977, 84, 110-115.
66. Takenaga, H.; Magaribuchi, T.; Nakajima, H. Japan. J. Pharm. 1979, 28, 457-464.
67. Holk, M.; Thorens, S.; Haeusler, G. European J. Pharm. In press, 1982.
68. Vogel, S.; Crampton, R.; Sperelakis, N. J. Pharm. Exp. Therap. 1979, 210, 378-385.
69. Patmore, L.; Whiting, R.L. Brit. J. Pharm. 1982, 75, 149P.
70. Ikeda, N.; Kodama, I.; Shibata, S.; Kondo, N.; Yamada, K. Cardiovasc. Pharm. 1982, 4, 70-75.

RECEIVED July 14, 1982.

2

Chemical Pharmacology of Calcium Antagonists

D. J. TRIGGLE

State University of New York, School of Pharmacy, Department of Biochemical Pharmacology, Buffalo, NY 14260

The comparative chemical pharmacology of a structurally diverse group of Ca^{2+} antagonists including verapamil/D600, diltiazem, nifedipine, cinnarizine, flunarizine and lidoflazine is reviewed. These agents, the Ca^{2+}-channel antagonists or Ca^{2+} entry blockers, appear to function by blocking Ca^{2+} entry through defined Ca^{2+} channels of the plasma membrane. The existence of structure-activity relationships, particularly for the verapamil/D600 and the nifedipine 1,4-dihydropyridine series, including stereoselectivity of action in smooth and cardiac muscle preparations is consistent with specific loci of action. Although these loci remain to be defined much evidence, including the demonstration of specific binding of ^{3}H-nitrendipine (a nifedipine analog) to smooth and cardiac muscle membranes, is indicative of a membrane site of action. Consistent with their chemical heterogeneity, the pharmacologic properties of these compounds are not uniform which suggests that a multiplicity of sites and mechanisms of action exist to impede Ca^{2+} entry. Of particular importance in this respect is the existence of tissue selectivity.

The group of compounds variously referred to as Ca^{2+} antagonists, Ca^{2+} channel , slow channel or Ca^{2+} entry blockers has achieved considerable pharmacologic and therapeutic attention. This group of compounds which includes verapamil, diltiazem, and nifedipine (Fig. 1) amongst its most prominent members was linked to Ca^{2+} metabolism in the pioneering work of Fleckenstein who showed that these agents mimicked in many respects the effects of Ca^{2+} withdrawal in smooth and cardiac muscle and represented a new pharmacologic approach to the control of excitation-contraction coupling in these tissues (1-3). Amongst

0097-6156/82/0201-0017$06.50/0

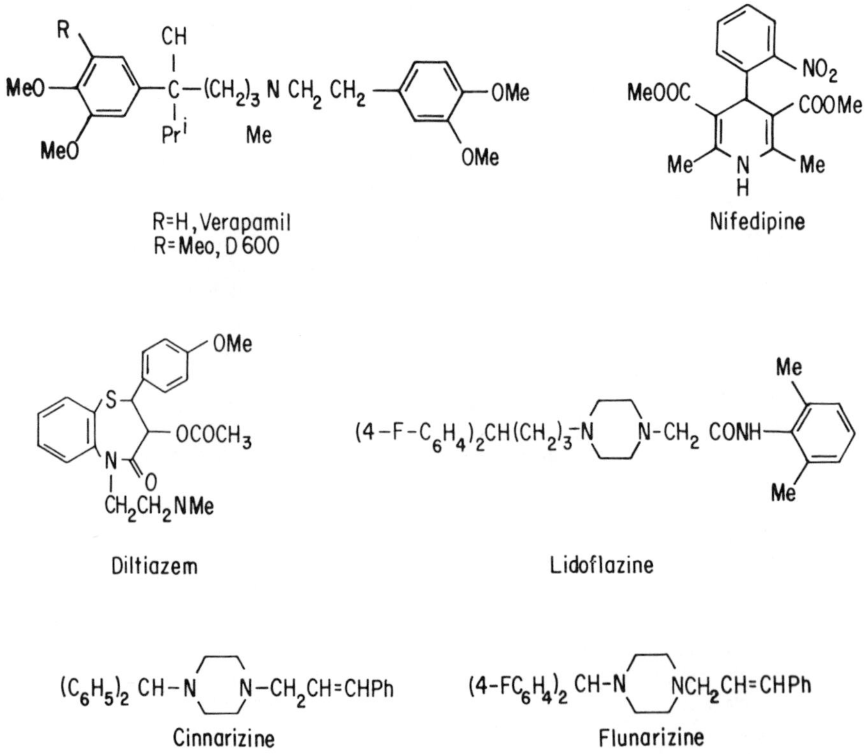

R=H, Verapamil
R=Meo, D 600

Nifedipine

Diltiazem

Lidoflazine

Cinnarizine

Flunarizine

Figure 1. Structural formulas of some Ca^{2+} antagonists.

the actual and potential therapeutic uses of these agents may be noted coronary vasodilation, cardiac arrhythmias, obstructive cardiomyopathies, hypertension, vasospastic disorders and myocardial preservation (4-6). Several recent reviews are available (4-9).

An understanding of the mechanism(s) of action of the Ca^{2+} channel antagonists must be set against the broader perspective of cellular Ca^{2+} regulation. In the resting cell the free intracellular Ca^{2+} level is maintained at $<10^{-7}$M and during stimulation it may rise to 10^{-7} - 10^{-5}M. The function of Ca^{2+} is to interact with Ca^{2+} binding proteins within the cell, including troponin-C and calmodulin, which serve to confer Ca^{2+} sensitivity to mechanical, secretory and enzymatic processes (10-12). Since there exists a very large driving force for Ca^{2+} entry into the cell, there must exist specific Ca^{2+} entry, exit, binding and sequestration mechanisms. These are shown in schematic form in Figure 2. The existence of these multiple sites serving to regulate Ca^{2+} concentration and function suggests that there may exist a corresponding multiplicity of agents serving to inhibit Ca^{2+} function (9,13,14). There is no lack of compounds possessing some degree of Ca^{2+}-antagonistic properties, although often secondary to other and better defined activities, and in only a few instances have the sites of Ca^{2+}-antagonist action been delineated.

A somewhat hetergenous group of compounds including trifluoperazine and some related antipsychotic agents, N-6-aminohexyl)-5-chloro-1-naphthalene sulfonamide (W-7) and local anesthetics function as calmodulin antagonists, apparently by interacting with a hydrophobic region on calmodulin exposed in the Ca^{2+}-calmodulin complex (15-17). It is clear, however, that many of these calmodulin antagonists exert other and better defined pharmacological actions at concentrations lower than those at which they inhibit calmodulin function.

The second major group of compounds, the Ca^{2+}-channel antagonists or Ca^{2+} entry blockers, are also a structurally heterogenous group of compounds and appear to function primarily by inhibiting Ca^{2+} uptake through defined pathways in the plasma membrane. These compounds will form the major focus of this chapter. Special, but not exclusive, emphasis will be placed on their interactions in smooth muscle.

Ca^{2+} Channels and Ca^{2+} Entry Processes

Ca^{2+} mobilized in response to membrane stimuli is derived from either intra- or extracellular sources (Figure 2). Membrane Ca^{2+} channels mediating Ca^{2+} entry have been classified into two major types (18-20). Receptor-operated channels (ROC, Figure 2) are associated with membrane receptors and are activated by specific agonist-receptor interactions, whilst potential-dependent channels (PDC, Figure 2) are activated by membrane depolariza-

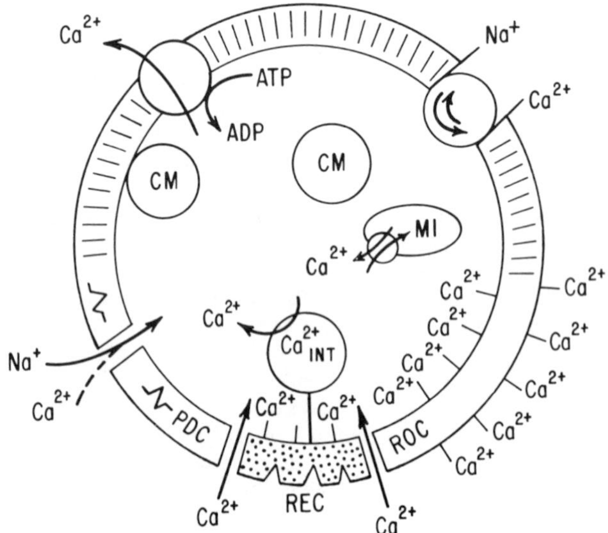

Figure 2. Schematic representation of cellular Ca^{2+} regulation.

In addition to a "leak pathway" (not shown), Ca^{2+} can enter the cell through several discrete channels as a minor component through the Na^+ channels and through the potential dependent (PDC) and receptor operated (ROC) Ca^{2+} channels. Ca^{2+} may also be released in response to a membrane signal from an intracellular Ca^{2+} store (Ca^{2+}_{INT}). Intracellular Ca^{2+} levels are regulated through the operation of membrane pumps, including Ca^{2+}–ATPase and a Na^+:Ca^{2+} exchange process. Additionally, intracellular organelles including mitochondria and sarcoplasmic reticulum can both sequestrate and release Ca^{2+}. The functions of intracellular Ca^{2+} are mediated through Ca^{2+}-binding proteins, a homologous group of proteins including calmodulin (CM) shown in cytosolic and membrane-associated states.

tion. Thus a solely depolarizing signal (elevated K^+) will activate only the PDC population whilst a chemical signal may activate ROC and PDC populations and can mobilize intracellular Ca^{2+}. The relative importance of these processes will be both tissue- and receptor-dependent.

The potential-dependent Ca^{2+} channels can be distinguished from other cation permeant channels by a variety of criteria including ion selectivity, voltage characteristics and pharmacologic sensitivity. The Ca^{2+} channel antagonists (Figure 1) exert their actions primarily at the potential-dependent Ca^{2+} channel. Electrophysiologic studies, almost exclusively in cardiac tissue, have demonstrated the ability of these agents to inhibit the slow inward (plateau) Ca^{2+} current (1-3,19,21). It is to be emphasized, however, that the sites of action of these agents remain to be defined precisely, that electrophysiologic studies are not available for most smooth muscle preparations and that the compounds are heterogeneous in both structure and activity. The term Ca^{2+} entry blockers may be preferable since it avoids any specific definition of site of action.

These agents are potent antagonists of K^+-induced responses in smooth muscle and are similarly effective against some agonist-induced responses (Table 1). Furthermore, their antagonist activity is dependent upon the level of extracellular Ca^{2+}, a competitive relationship having been described in some systems (13,19,22,23) and they have been shown to inhibit Ca^{2+} uptake into a number of smooth muscles at the same concentration ranges at which they inhibit mechanical responses (9,13,24-26). These studies indicate that the Ca^{2+} entry blockers, including verapamil, D600, nifedipine, diltiazem and cinnarizine are potent antagonists of Ca^{2+} movements through one class of Ca^{2+} channels. These studies alone, however, indicate neither the sites nor the mechanisms of action of these compounds.

Structure-Activity Relationships and Sites of Action

A particularly obvious characteristic of the Ca^{2+} entry blockers is their diversity of structure. This diversity makes it quite unlikely that a single inclusive structure-activity relationship exists for this class of compounds. Consequently it is probable that there exist multiple modes and sites of action (9,13,14,31).

Some aspects of the pharmacologic activity of these compounds are quite consistent with this conclusion. The negative inotropic and Ca^{2+}-current blocking properties of verapamil and D600 are frequency- and voltage-dependent (32-35). Potency increases with increasing frequency of stimulation of cardiac tissue. One explanation of such findings is that verapamil and D600 preferentially block the Ca^{2+} channel subsequent to its activation. Activity is also dependent upon membrane potential, increasing with decreasing membrane potential (less negative) and verapamil and D600 impede the voltage-dependent recovery of the Ca^{2+} channel from inactivation. Thus, blockade of Ca^{2+} channel function by verapamil or D600 is a state-dependent process, removal of blockade occurring most rapidly from the resting channel state.

Similar conclusions have been advanced previously to accomodate the voltage-dependent inhibition of Na^{+} channels by local anesthetics (36,37). In contrast, nifedipine lacks almost completely such behaviour suggesting a major difference in behaviour to that of verapamil/D600 (38,40).

The extent to which the above conclusions may extend to smooth muscle is not clear. However, a "use-dependence" of flunarizine action has been described in depolarized vascular smooth muscle (41,42).

As yet there are few quantitative structure-activity studies for the Ca^{2+} channel antagonists. However, qualitative indices of structure-function dependence are available, particularly for the 1,4-dihydropryidines of the nifedipine class where, because of the comparative ease of synthesis (43), a large number of analogs are available.

The general structural requirements for activity in the 1,4-dihydropyridine series are shown in Figure 3. Quite generally the following observations hold (43-49):

1. N-1 of the 1,4-dihydropyridine ring must be unsaturated (R^{0} = H).
2. The 2,6-substituents (R^{2},R^{6}) of the 1,4-dihydropyridine ring should be lower alkyl groups, although one NH_2 group appears to be acceptable.
3. Ester functions (R^{3},R^{5}) are optimum for the 3- and 5-positions of the 1,4-dihydropyridine ring. Replacement by other groups including -CN or -COMe leads to substantial loss of activity. However, there is considerable tolerance for the nature of the ester functions, quite bulky groups maintaining or even increasing activity relative to the carbomethoxy functions of nifedipine.
4. An aryl substituent (R^{4}), preferably substituted phenyl, appears optimum in the 4-position of the 1,4-dihydropyridine ring. The position of the substituent in the phenyl ring is quite critical: 4'-substitution is invariably highly detrimental, whereas 2'- and 3'-substitutions increase activity.
5. The 1,4-dihydropyridine ring is essential. Conversion to the oxidized pyridine or the fully reduced piperidine structures abolishes activity.

It is of interest that similar structural requirements appear to hold for vasodilation, negative inotropic activity and smooth muscle relaxation. This suggests a qualitative similarity

TABLE I Ca^{2+}-Channel Antagonist Activities in Smooth Muscle Preparations (9, 13)

SYSTEM	ANTAGONIST		ID_{50},M	Ref.
Rabbit aorta	NE	Verapamil	1.2×10^{-4}	27
	K^+	Verapamil	2.7×10^{-8}	
Canine trachea	ACh	Verapamil	$> 10^{-4}$	28
	K^+	Verapamil	$<< 10^{-4}$	
Rabbit saphenous artery	5-HT	Nimodipine	$> 2 \times 10^{-5}$	29
	K^+	Nimodipine	2.5×10^{-10}	
Rabbit basilar artery	5-HT	Nimodipine	7.3×10^{-10}	29
	K^+	Nimodipine	1.7×10^{-10}	
Canine coronary	NE	D600	5×10^{-7}	30
	K^+	D600	2×10^{-7}	
G.P. ileum	ACh	Nifedipine	5.1×10^{-9}	26
	K^+	Nifedipine	3.4×10^{-9}	

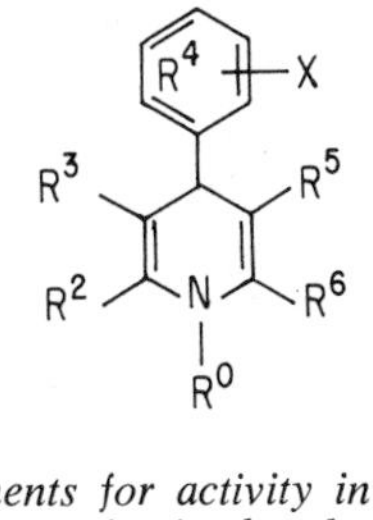

Figure 3. Structural requirements for activity in the 1,4-dihydropyridine series. See text for further details.

of the sites of 1,4-dihydropyridine interaction in these several preparations.

A quantitative structure-activity relationship for the negative inotropic activity of a small series of 1,4-dihydropyridines has been developed (47). For 2,6-dimethyl-3,5-dicarbomethoxy-4-(substituted phenyl)-1,4-dihydropyridines the effect of the phenyl substituent is determined primarily by steric effects. Thus, for ortho-substituted derivatives,

$$\log \frac{1}{EC_{50}} = 5.06 + 0.80\, B_1 \quad (n = 8,\ r = 0.91)$$

where B_1 is the minimum width of the substituent.

The importance of steric factors in determining the activity of 1,4-dihydropyridines is also suggested from the solid state structures of a series of 1,4-dihydropyridines (50,51). The 4-phenyl ring in these compounds (I) is oriented approximately bisecting the 1,4-dihydropyridine ring and the latter deviates from planarity, N_1 and C_4 forming the apices of a boat conformation. A good correlation exists between the planarity of the 1,4-dihydropyridine ring and pharmacologic activity (Fig. 4), such that activity decreases with decreasing ring planarity.

In the dihydropyridine class it is also evident that the ester functions at C_3 and C_5 of the 1,4-dihydropyridine ring are an important determinant of activity. Rodenkirchen *et al* (47), studying a limited series of compounds producing negative inotropic effects in cardiac muscle, suggested that activity decreases with increasing size of lipophilicity of the ester function. However, other studies (48,49) have shown that bulky ester substitutions maintain or even increase activity in smooth muscle preparations (Table 2).

Structure-activity data are quite restricted for other classes of Ca^{2+} channel antagonists. For a limited series of verapamil derivatives carrying ring substituents (II), negative inotropic activity depends upon both electronic and steric properties of the substituent (52,53),

$$\log \frac{1}{EC_{50}} = 0.93\, F - 0.59\, MR \quad (n = 13,\ r = 0.82)$$

substituents reducing electron density by an inductive mechanism increase activity. Inductive effects will be more powerful for meta than for para-substituents, hence D600 is more potent than verapamil.

The existence of such structure-activity relationships is consistent with a specific, rather than a non-specific, site of action. Additional support is provided by the stereoselectivity of action exhibited by verapamil/D600 and by several agents of the 1,4-dihydropyridine class.

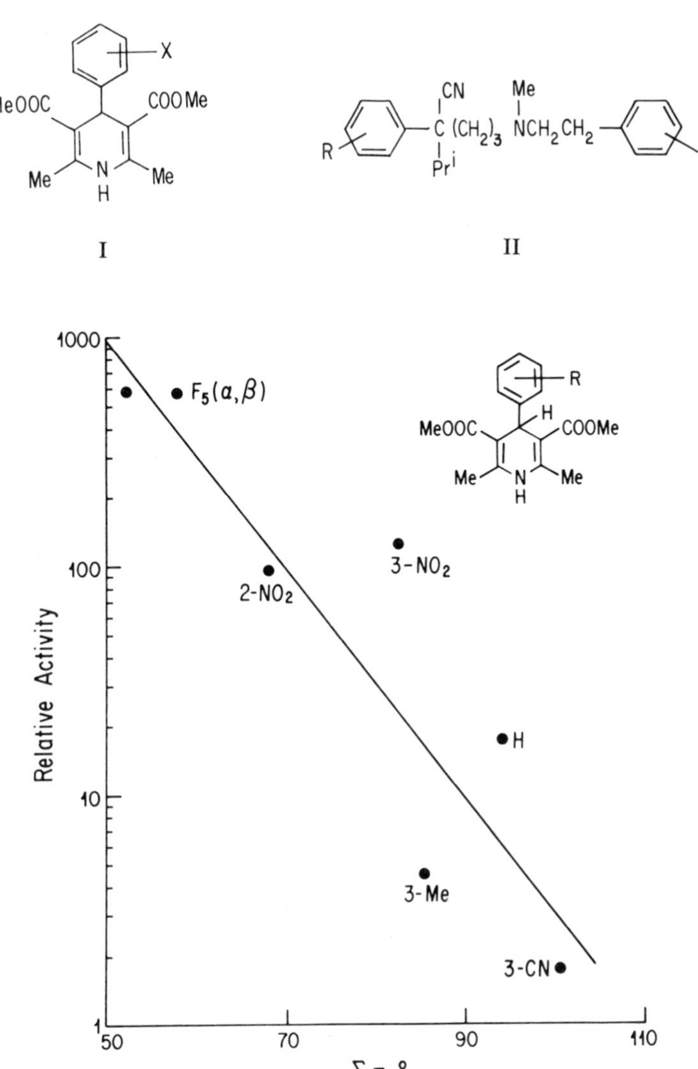

Figure 4. Relationship between relative pharmacologic activity (muscarinic-induced Ca^{2+}-dependent responses of guinea pig ileal longitudinal muscle; nifedipine, 2-NO_2 = 100) and degree of ring puckering, $\Sigma\tau^0$ (defined as the sum of the absolute values of the six ring torsion angles) for a series of 1,4-dihydropyridines. Data from Ref. 51.

TABLE II Pharmacologic Activities of Ester-Substituted Analogs of Nifedipine (63,64)

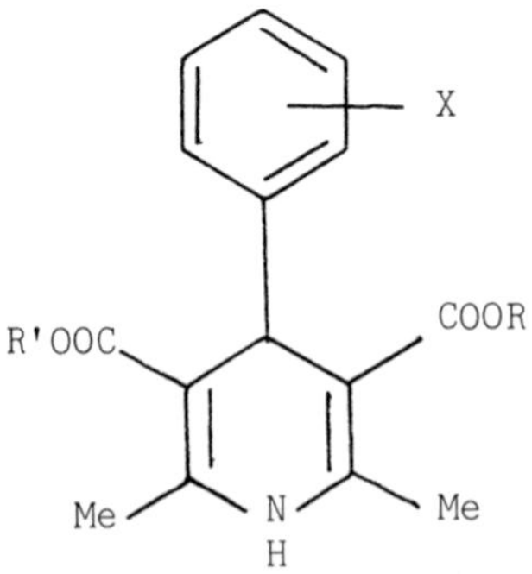

X
R'OOC
COOR
Me
N
H
Me

X	R	R'	System ID_{50}, M G.P. ileum Ba^{2+}	G.P. ileum K^+	Rabbit aorta K^+
$2\text{-}NO_2$	Me	Me (Nifedipine)	1.4×10^{-8}	2.6×10^{-10}	8.1×10^{-9}
$3\text{-}NO_2$	Me	Me	2.3×10^{-8}	8.9×10^{-10}	-
$3\text{-}NO_2$	Me	Et (Nitrendipine)	-	4.6×10^{-10}	-
$3\text{-}NO_2$	Me	CH_2CHMe_2 (Nisoldipine)	-	3.9×10^{-10}	-
$3\text{-}NO_2$	$CHMe_2$	CH_2CH_2OMe (Nimodipine)	-	4.2×10^{-10}	-
$3\text{-}NO_2$	Me	$CHMe_2$(-)	1.5×10^{-9}	-	5×10^{-10}

Stereoselectivity of verapamil and D600 actions has been reported for both cardiac and smooth muscle. The S (-) enantiomers are the more potent negative inotropic agents, Ca^{2+} current antagonists and smooth muscle relaxants (9,13,54-58). Although the ratios reported vary considerably (Table 3), stereoselectivity is seen consistently in the inhibition of Ca^{2+}-dependent events and quite generally is absent in non-Ca^{2+} events (58,60). Stereoselectivity has also been reported for 1,4-dihydropyridines bearing non-identical ester substituents at the 3- and 5-positions (Table 4).

Although structure-activity data provide evidence for specificity of action, they do not define the sites of action of the Ca^{2+} antagonists. Increasing evidence points, however, to a membrane locus of action. D-600 and diltiazem have been shown to be ineffective in skinned cardiac and smooth muscles (19,65), compatible with a plasma membrane site of action. Consistent with this, high concentrations of Ca^{2+} channel antagonists are necessary to inhibit intracellular Ca^{2+} movements (9,13,66,67). Additional evidence for a plasma membrane site of action comes from recent studies describing the specific binding of ^{3}H-nitrendipine (2,6-dimethyl-3-carbomethoxy-5-carboethoxy-4-(3-nitrophenyl)-1,4-dihydropyridine) to cardiac and smooth muscle membranes (68-70). Binding is saturable, reversible and a good correlation exists between the Ca^{2+}-antagonist activities of a series of 1,4-dihydropyridines and their ability to compete with ^{3}H-nitrendipine binding (Figure 5). These studies indicate very clearly that the binding site for nitrendipine is that at which pharmacological response is controlled. However, the specific nature of this site remains to be determined.

Selectivity of Action

The Ca^{2+}-entry blockers are heterogeneous with respect both to structure and pharmacologic activity. They differ substantially with respect to their cardiac depressant and vasodilatory properties, their activities on different vascular tissues or beds, their onset times and durations of action, their use dependence, the extent to which they inhibit stimulus-secretion coupling processes and the extent to which they show antagonism of events other than Ca^{2+} entry (9,13,19,71,72). These differences suggest selectivity of action, although it is likely that such selectivity has several origins.

The data of Table 1 reveal very clearly that some tissues (aorta, trachea) show a clear distinction in sensitivity between agonist- and K^{+}-induced responses whilst other tissues (ileum) show equisensitivity of K^{+}-and agonist-induced responses. Similarly, marked differences in sensitivity to responses induced by the same agonist in different tissues are found (5-HT responses of rabbit saphenous and basilar arteries).

TABLE III Stereoselectivity of Verapamil and D600 Action

SYSTEM	Verapamil Ratio	D600 Ratio	Ref.
CARDIAC MUSCLE	(-) : (+)	(-) : (+)	
Canine heart, A.V. block	10		55
negative chronotropic	3		55
Cat heart, papillary muscle	10	100	56-58
Dog heart, negative inotropic	15		
negative chronotropic	5		59
SMOOTH MUSCLE			
Guinea-pig ileum, ACh		180	60
Ca^{2+}/K^{+}		10	60
Rat portal vein, NE		25	60
Rat vas deferens, Ca^{2+}/K^{+}		40	60
Rabbit aorta, Ca^{2+}/K^{+}	10	60	61

TABLE IV Stereoselectivity of
1,4-Dihydropyridine Action (<u>62</u>-<u>64</u>)

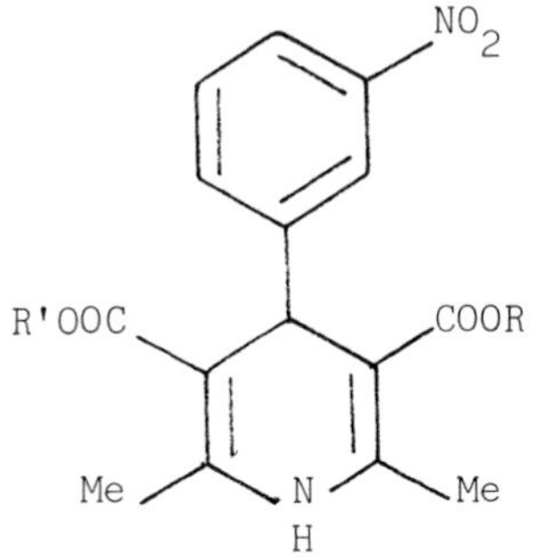

NO_2

R'OOC COOR

Me N Me
H

ID_{50},M : Isomer Ratio

R	R'	G.P. Ileum/K^+		G.P. Bladder/K^+		Rabbit Aorta/K^+	
Me	$CH_2CH_2NMeCH_2Ph$						
	(+)	$3.1x10^{-10}$	4	-		-	
	(-)	$1.3x10^{-9}$					
Me	CH Me_2						
	(+)	$4.1x10^{-9}$	10	$4.0x10^{-9}$	8	$1.4x10^{-8}$	10
	(-)	$4.2x10^{-10}$		$5.6x10^{-10}$		$1.5x10^{-9}$	
Et	CH Me_2						
	(+)	$4.9x10^{-9}$	2.5	-		-	
	(-)	$1.9x10^{-9}$					

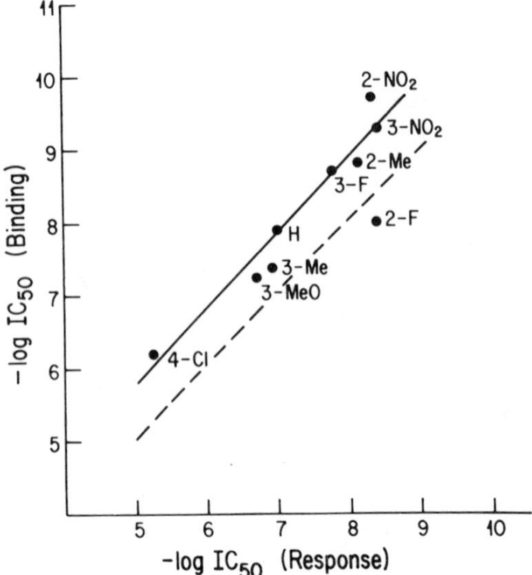

Figure 5. Correlation between affinities of a series of 2,6-dimethyl-3,5-dicarbomethoxy-4-substituted phenyl-1,4-dihydropyridines measured as their abilities to displace specific ^3H-nitrendipine binding in a microsomal preparation of guinea pig ileal longitudinal smooth muscle (IC_{50}, binding) and to inhibit by 50% the sustained mechanical response of the intact muscle to muscarinic receptor stimulation (IC_{50}, response). Key: ——, best fit line for data and – – –, anticipated line for 1:1 correlation.

The negative inotropic effects of D600 on atria and papillary muscles lie in the order rabbit > guinea pig > rat consistent with the relative dependence of response upon extracellular Ca^{2+} (73). Within a single tissue marked differences in sensitivity to different agonists may exist as, for example, in the canine trachea where the order of sensitivity of responses to D600 is, K^+ > 5-HT > Hist > ACh (28,74). Finally, differences in antagonist sensitivity may occur with activation of different receptor subpopulations in the same tissue. In vascular smooth muscle, where postsynaptic α_1, - and α_2-receptors occur, the antagonists have been variously reported to discriminate against α_1- or α_2-induced mechanical responses in *in vitro*, arterial and venous vascular preparations (75-77). These differences probably reflect the varying degree to which receptor and channel activation processes are linked and this may differ significantly depending upon agonist, tissue and species.

In addition to the preceding evidence that the extent of channel activation may differ there are several lines of evidence indicating that differences in the sites or mechanisms of antagonist action exist between tissues. The relative cardiac: smooth muscle selectivity clearly differs significantly amongst the several types of antagonist. Verapamil and D600 exhibit the least selectivity, and smooth muscle selectivity is most pronounced with nifedipine, flunarizine and lidoflazine (9,13,71,72, 78-82). Of particular interest are substantial indications that some antagonists show selectivity for particular vascular tissues or beds. Thus, myogenic tone in vascular tissue is sensitive to nifedipine and D600 but is quite insensitive to flunarizine and lidoflazine. In a comparison of Ca^{2+}-induced responses in depolarized vascular smooth muscle van Neuten and Vanhoutte have shown that the range of ED_{50}'s for flunarizine exceeds 200 (70). Tiapamil (Ro 11-1781, N-(3,4-dimethoxyphenethyl)-2-(3,4-dimethoxyphenyl) - N-methyl-m-dithiane-2-propylamine-1,1,3,3-tetraoxide), an analog of verapamil, provides a further example of tissue selectivity (83,84). Tiapamil has apparent selectivity, with respect to verapamil for the coronary vasculature and unlike verapamil appears to be inactive in non-vascular smooth muscle.

In addition to such selectivity differences marked differences exist among the various antagonists in their time course of action. In particular, cinnarizine, flunarizine and lidoflazine are very slow to reach their maximum antagonism, requiring some 30-60 minutes. In this respect these compounds differ markedly from verapamil, nifedipine and diltiazem.

The majority of studies with Ca^{2+} antagonists have focussed on cardiac and smooth muscles. However, several studies are available with stimulus-secretion coupling systems (for reviews see references 9,13,85,86). Particularly for neuronal systems (neurotransmitter release) these systems appear markedly less sensitive to the Ca^{2+} entry blockers than cardiac or smooth muscle.

Although a great deal of comparative work is needed, the evidence thus far available indicates that substantial tissue selectivity can occur with a number of Ca^{2+} entry blockers. The mechanism(s) underlying this selectivity are not established. It is entirely possible, however, that differences in Ca^{2+} channel structure and kinetics are responsible for this selectivity.

Other Sites Of Action

Although the Ca^{2+}-entry blockers are, with justification, known to inhibit specific Ca^{2+} mobilization processes they can inhibit other events including ion movements and receptor binding (9,13,60). Such additional properties may contribute to the spectrum of pharmacologic and therapeutic activities of these agents.

Verapamil and D600 have been shown to inhibit competitively α_1 and α_2-adrenergic, muscarinic and opiate receptor binding with K_I values in the range 10^{-6} - 10^{-5}M (Table 5). Although these activities are close to those used in many studies as indicative of Ca^{2+}-antagonistic effects some discrimination may be possible, since inhibition by D600 of receptor binding is non-stereoselective (60). Additionally D600 and verapamil have been shown to inhibit Na^+ and K^+ currents in several excitable preparations with IC_{50} values of 10^{-6} - 10^{-5}M and the post-synaptic membrane currents induced by acetylcholine (Table 5; 94-96).

TABLE V Other Effects of Ca^{2+}-entry Blockers

A. Inhibition of receptor binding

Ant.	Receptor	Ligand	K_I,M	Ref.
(±)-D600	Musc. (g.p. ileum)	^{3}H-QNB	6.9×10^{-6}	60
(+)-D600	Musc. (g.p. ileum)	^{3}H-QNB	8.4×10^{-6}	60
(-)-D600	Musc. (g.p. ileum)	^{3}H-QNB	7.0×10^{-6}	60
(±)-D600	Musc. (g.p. ileum)	^{3}H-QNB	8.8×10^{-6}	87
(±)-D600	α_1-Adr. (rat brain)	^{3}H-WB4101	1.1×10^{-6}	87
(±)-D600	α_1-Adr. (rat brain)	^{3}H-WB4101	0.9×10^{-6}	88
(±)-D600	α_1-Adr. (rat vas. def.)	^{3}H-WB4101	2.2×10^{-6}	60
(+)-D600	α_1-Adr. (rat vas. def.)	^{3}H-WB4101	1.5×10^{-6}	60
(-)-D600	α_1-Adr. (rat vas. def.)	^{3}H-WB4101	1.7×10^{-6}	60
(±)-D600	Opiate (rat brain)	^{3}H-Naloxone	4.0×10^{-6}	87
(±)-D600	α_2-Adr. (rat brain)	^{3}H-Clonidine	9.3×10^{-6}	89
(±)-D600	α_1-Adr. (rat brain)	^{3}H-Prazosin	2.8×10^{-6}	89

B. Inhibition of ionic events

Ant.	System	IC_{50},M	Ref.
(±)-D600	Veratridine-stim. Na^+-uptake (heart, neuroblastoma)	3×10^{-6}	90
(±)-D600	Veratridine depol. synaptosome	5×10^{-6}	91
(±)-D600	Na^+, K^+ current, Cardiac purkinje	$\sim 1 \times 10^{-6}$	92
(±)-Verapamil	K^+, pancreatic islet cells	$\sim 10^{-5}$	93
(±)-D600	Cl^-, Na^+, K^+-ACh, Mollusc neurons	-	94
(±)-D600,diltiazem	Na^+, K^+-ACh, frog end plate	-	95,96

Acknowledgments

Preparation of this manuscript was supported by a grant from the National Institutes of Health (HL 16003). Additional support from the Miles Institute for Preclinical Pharmacology is gratefully acknowledged.

Literature Cited

1. Fleckenstein, A. Verh. Dtsch. Ges. Inn. Med., 1964, 70, 81-89.
2. Fleckenstein, A.; Tritthart, A.,; Fleckenstein, B.; Herbst, A.; Grun, G. Pflug. Arch. ges. Physiol. 1969, 307, R25.
3. Fleckenstein, A.: "New Perspectives on Calcium Antagonists", Weiss, G.B., Ed.; Amer. Physiol. Soc. Bethesda, MD, 1981, p. 59.
4. Ellrodt, G.; Chew, C.Y.C.; Singh, B.N. Circulation 1980, 62, 669-679.
5. Antman, E.M.; Stone, P.H.; Muller, J.E.; Braunwald, E. Ann. Int. Med. 1980, 93, 875-885.
6. Stone, P.H.; Antman, E.M.; Muller, J.E.; Braunwald, E. Ann. Int. Med. 1980, 93, 836-904.
7. Zelis, R.F.; Schroeder, J.S. (Eds.). Calcium, Calcium Antagonists and Cardiovascular Disease, Chest (Supplement) 1980, 78, 121-247.
8. Weiss, G.B. (Ed.), New Perspectives on Calcium Antagonists, Amer. Physiol. Soc. Bethesda, MD, 1981.
9. Triggle, D.J.; Swamy, V.C., Circ. Res., in press (1982).
10. Kretsinger, R.H., Adv. Cycl. Nuc. Res. 1979, 11, 1-26.
11. Means, A.R.; Dedman, J.R., Nature, 1980, 285, 73-77.
12. Cheung, W.Y. Science, 1980, 207, 19-24.
13. Triggle, D.J. "New Perspectives on Calcium Antagonists", Weiss, G.B., Ed. Amer. Physiol. Soc. Bethesda, MD, 1981, p. 1.
14. Triggle, D.J.; Swamy, V.C., Chest (Supplement), 1980, 78, 174-179.
15. Levin, R.M.; Weiss, B., J. Pharmacol. Exptl. Therap. 1979, 208, 454-459.
16. Kanamori, M.; Naka, M.; Asano, M.; Hidaka, H., J. Pharmacol. Exptl. Therap. 1981, 217, 494-499.
17. Roufogalis, B.: "Calcium and Cell Function", Vol. 3. Cheung, W.Y., Ed., Academic, London and New York, in press.
18. Bolton, T.B., Physiol. Rev. 1979, 59, 606-718.
19. Fleckenstein, A., Ann. Rev. Pharmacol. Toxicol. 1977, 17, 149-166.
20. van Breemen, C.; Aaronson, P.; Loutzenhiser, R.; Meisheri, K. Chest (Suppl.) 1980, 78, 157-165.
21. Bayer, R.; Ehara, T. Prog. Pharmacol. 1978, 2, 31-37.
22. Imai, S. Trends Pharmacol. Sci. 1979, 2, 87-89.

23. Jim, K.; Harris, A.; Rosenberger, L.B.; Triggle, D.J. Eur. J. Pharmacol. 1981 76, 67-72.
24. Meisheri, K.D.; Hwang, O.; van Breemen, C. J. Mem. Biol. 1981, 59, 19-25.
25. Godfraind, T.; Dieu, D. J. Pharmacol. Exptl. Therap. 1981 217, 510-515.
26. Rosenberger, L.B.; Ticku, M.K.; Triggle, D.J.; Can. J. Physiol. Pharmacol. 1979, 289, 333-347.
27. Schümann, H.J.; Gorlitz, B.D.; Wagner, J. Naunyn-Schmied. Arch. Pharmacol. 1975, 289, 409-417.
28. Farley, J.M.; Miles, P.R., J. Pharmacol. Exptl. Therap. 1978, 207, 340-346.
29. Towart, R. Circ. Res. 1981, 48, 650-657.
30. van Breemen, C.; Siegel, B. Circ. Res. 1980, 46, 426-429.
31. Henry, P.D. Pract. Cardiol. 1979, 5, 145-156.
32. Bayer, R.; Hennekes, R.; Kaufmann, R.; Mannhold, R. Naunyn-Schmied. Arch. Pharmacol. 1975, 290, 49-68.
33. Ehara, T.; Kaufmann, R. J. Pharmacol. Exptl. Therap. 1978, 207, 49-55.
34. Kohlhardt, M. Basic Res. Cardiol. 1981, 76, 589-601.
35. McDonald, T.F.; Pelzer, D.; Trautwein, W. 1980, Pflüg. Arch. Physiol. 385, 175-179.
36. Hondagheim, L.M.; Katzung, B.G., Biochim. Biophys. Acta, 1977, 472, 373-398.
37. Kendis, J.J.; Courtney, R.R.; Cohen, E.N. J. Pharmacol. Exptl. Therap. 1979, 210, 446-452.
38. Kohlhardt, M.; Fleckenstein, A. Naunyn-Schmied. Arch. Pharmacol. 1977, 298, 267-272.
39. Bayer, R.; Ehara, T. Prog. Pharmacol. 1978, 2, 31-37.
40. Kohlhardt, M; Haap, K. Naunyn-Schmied. Arch. Pharmacol.1981, 316, 178-185.
41. Godfraind, T.; Dieu, D. J. Pharmacol. Exptl. Therap. 1981, 217, 510-515.
42. Godfraind, T.; Miller, R.C. Brit. J. Pharmacol. 1982, 75, 229-236.
43. Bossert, F.; Meyer, H.; Wehinger, E. Angewandte Chemie. Int. Ed. 1981, 20, 762-769.
44. Loev, B.; Goodman, M.M.; Snader, K.M.; Tedeshchi, R.; Macko, E. J. Med. Chem. 1974, 17, 956-965.
45. Iwanami, M.; Shibanuma, T.; Fujimoto, M.; Kawai, R.; Tamazawa, K.; Takenaka, T.; Takahashi, K.; Murakami, M. Chem. Pharm. Bull. Tokyo, 1979, 27, 1426-1440.
46. Rosenberger, L.B.; Triggle, D.J. "Calcium and Drug Action", G. Weiss, G.B., Ed. Plenum; 1978, New York, p.3.
47. Rodenkirchen, R.; Bayer, R.; Steiner, R.; Bossert, F.; Meyer, H.; Moller, E., 1979, Naunyn-Schmied. Arch. Pharmacol. 1979, 310, 69-78.
48. Bossert, F.; Horstmann, H.; Meyer, H.; Vater, W., Arz. Forsch. 1979, 29, 226-229.

49. Meyer, H.; Bossert, F.; Wehinger, E.; Stoepel, K.; Vater, W. Arz. Forsch. 1981, 31, 407-409,
50. Triggle, A.M.; Shefter, E.; Triggle, D.J., J. Med. Chem. 1980, 23, 1442-1444.
51. Fossheim, R.; Svarteng, K.; Mostad, A.; Rømming, C.; Shefter, E.; Triggle, D.J. J. Med. Chem. 1982, 25, 126-131.
52. Mannhold, R.; Steiner, R.; Haas, W.; Kaufmann, R. Naunyn-Schmied. Arch. Pharmacol. 1978, 302, 217-226.
53. Mannhold, R.; Zierden, P.; Bayer, R.; Rodenkirchen, R.; Steiner, R., Arz. Forsch. 1981, 31, 773-780.
54. Bayer, R.; Kaufmann, R.; Mannhold, R. Naunyn-Schmied. Arch. Pharmacol. 1975, 290, 69-80.
55. Kaumann, A.J.; Serur, J.R. Naunyn-Schmied. Arch. Pharmacol. 1975, 291, 347-358.
56. Kaufmann, R.; Bayer, R.; Hennekes, R.; Kalusche, D.; Mannhold, R.; Naunyn-Schmied. Arch. Pharmacol. 1974, 285, R39.
57. Ludwig, C.; Nawrath, H., Brit. J. Pharmacol. 1977, 59, 411-417.
58. Nawrath, H.; Blei, I.; Gegner, R.; Ludwig, C.; Zang, X-G. "Calcium, Antagonists in Cardiovascular Therapy", Zanchetti, A., Krikler, D.M. Eds., Excerpta Med. Amsterdam, 1981, p. 52.
59. Satoh, K.; Yanagisawa, T.; Taira, N., J. Cardiovas. Pharmacol. 1980, 2, 309-318.
60. Jim, K.; Harris, A.; Rosenberger, L.B.; Triggle, D.J., Eur. J. Pharmacol. 1981, 76, 67-72.
61. Raschack, M.; Engelman, K. International Society for Heart Research. 10th International Congress, Moscow, USSR, September 1980 Abs.
62. Shibanuma, T.; Iwanani, M; Okuda, K.; Takenaba, T.; Murakami, M. Chem. Pharm.Bull. 1980, 28, 2809-2812.
63. Towart, R.; Wehinger, E.; Meyer, H. Naunyn-Schmied. Arch. Pharmacol. 1981, 317, 183-185.
64. Triggle, D.J.; unpub. data.
65. Itoh, T.; Kajiwara, M.; Kitamura, K.; Kuriyama, H. Brit. J. Pharmacol. 1981, 74, 455-468.
66. Entman, M.L.; Allen, J.C.; Bornet, E.P.; Gillette, P.C.; Wallick, E.T.; Schwartz, A., J. Mol. Cell. Cardiol. 1972, 4, 681-687.
67. Thorens, S.; Haeusler, G. Eur. J. Pharmacol. 1979, 54, 79-91.
68. Bellemann, P.; Ferry, D.; Lübbecke, F.; Glossmann, H. Arz. Forsch. 1981, 31, 2064-2067.
69. Bolger, G.T.; Gengo, P.J.; Luchowski, E.M.; Siegel, H.; Triggle, D.J.; Janis, R.A. Biochem. Biophys. Res. Comm. 1982, 104, 1604-1609.
70. Ehlert, F.J.; Itoga, E.; Roeske, W.R.; Yamamura, H.I., Biochem. Biophys. Res. Comm. 1982, 104, 937-943.
71. Henry, P.D., Amer. J. Cardiol. 1980, 46, 1047-1058.
72. Vanhoutte, P.M., Circulation, 1982, 65 (Suppl. I), I, 11-19.

73. Siegl, P.K.S.; McNeill, J.H., Can. J. Physiol. Pharmacol. 1980, 58, 1406-1411.
74. Coburn, R.F., Fed. Proc. 1977, 36, 2692-2697.
75. Van Meel, J.C.A.; DeJonge, A.; Kalkman, H.O.; Wilffert, B.; Timmermans, P.B.M.W.M.; Van Zwieten, P.A., Eur. J. Pharmacol., 1981, 69, 205-208.
76. DeMey, J.; Vanhoutte, P.M., Circ. Res. 1981, 48, 875-884.
77. Nghiem, C.; Swamy, V.C.; Triggle, D.J., Life Sciences, 1982, 30, 45-49.
78. Hashimoto, K.; Takeda, K.; Katano, Y.; Nakagawa, Y.; Tsukada, T.; Hashimoto, T.; Shimamoto, N.; Sakai, K.; Otorrii, T.; Imai, S., Arzneim. Forsch., 1979, 29, 1368-1373.
79. VanNueten, J.M.; Vanhoutte, P.M.; Angiology, 1981, 32, 476-484.
80. Quinn, P.; Briscoe, M.G.; Nuttall, A.; Smith, H.J., Cardiovas. Res., 1981, 15, 398-403.
81. Godfraind, T.; Kaba, A.; Polster, P., Arch. Int. Pharmacodyn., 1968, 172, 235-239.
82. Godfraind, T., "New Perspectives on Calcium Antagonists", Weiss, G.B., Ed.; Amer. Physiol. Soc., Bethesda, MD., 1981, p. 95.
83. Eigenmann, R.; Blaber, L.; Nakamura, K.; Thorens, S.; Haeusler, G., Arz. Forsch. 1981, 31, 1393-1401.
84. Eigenmann, R.; Gerold, M.; Hefti, F.; Jovanovic, D.; Haeusler, G., Arz. Forsch. 1981, 31, 1401-1410.
85. Rubin, R.P., "New Perspectives on Calcium Antagonists", Weiss, G.B., Ed., Amer. Physiol. Soc., Bethesda, MD., 1981, p. 147.
86. Fleischer, N.; Schubart, U.K.; Fleckman, A.; Erlichman, J., "New Perspectives on Calcium Antagonists", Weiss, G.B., Ed., Amer. Physiol. Soc., Bethesda, MD., 1981, p. 177.
87. Fairhurst, A.S.; Whittaker, M.L.; Ehlert, F.J., Biochem. Pharmacol. 1980, 29, 155-162.
88. Atlas, D.; Adler, M., Proc. Nat. Acad. Sci., U.S.A., 1981, 78, 1227-1241.
89. Glossman, H.; Hornung, R. Mol. Cell. Endocrinol., 1980, 19, 243-251.
90. Galper, J.B.; Catterall, W. Mol. Pharmacol., 1979, 15, 174-178.
91. Nachsen, D.A.; Blaustein, M.P., Mol. Pharmacol., 1979, 15, 174-178.
92. Kass, R.S.; Tsien, R.W., J. Gen. Physiol. 1975, 66, 169-182.
93. Lebrun, P.; Malaisse, W.J.; Herchuelz, H. Biochem. Pharmacol. 1981, 30, 3291-3294.
94. Brgestovski, P.D.; Iljin, V.I., J. Physiol. Paris. 1980, 76, 515-522.
95. Miledi, R.; Parker, I., Proc. Roy. Soc. Lond. B., 1980, 211, 143-150.
96. Miledi, R.; Parker, I., Biomed. Res., 1981, 2, 587-589.

RECEIVED June 10, 1982.

3

Effects of Calcium Inhibitory Compounds upon the Cardiovascular System

WILLIAM W. MUIR

Ohio State University, Department of Veterinary Physiology and Physiology, Columbus, OH 43210

A diverse group of heterogeneous compounds have emerged which are advocated for the therapy of a variety of cardiovascular diseases including angina pectoris, hypertension, cardiac arrhythmias, obstructive forms of cardiomyopathy and ischemic heart disease. The therapeutic success of these compounds centers, in part if not totally, upon their ability to interfere with calcium metabolism. Drugs exhibiting the ability to inhibit the effects of calcium upon cardiovascular function have been classified as calcium antagonists, implying a receptor agonist-antagonist relationship. Verapamil, nifedipine, diltiazem and perhexiline are the most extensively studied of these compounds. The recent development of a new group of calcium inhibitory compounds, the 2-substituted 3-dimethylamino-5-,6-methylenedioxyindenes, are believed to inhibit the intracellular regulatory protein calmodulin.

Despite difficulties in determining the precise mechanism of action of drugs possessing calcium inhibitory activity, these compounds exert potent negative inotropic, chronotropic and dromotropic effects upon cardiac tissues. They are excellent coronary and peripheral vasodilators and preserve myocardial metabolism while preventing mitochondrial swelling during cardiac ischemia. Their antiarrhythmic activity in intact animals and isolated tissue preparations is closely linked to their ability to inhibit the slow inward calcium current, interfere with transmembrane sodium ion flux, and indirect effects.

The development of specifically designed and precise experimental techniques coupled with an increased understanding

0097-6156/82/0201-0039$09.50/0

of the changes that occur in myocardial metabolism during disease have been instrumental in advancing cardiac pharmacotherapeutics. Nowhere in the science of pharmacology as it relates to cardiology has this development been made more evident than in the synthesis and clinical usage of adrenergic antagonists and a group of compounds collectively termed "calcium antagonists". The latter compounds have derived their name primarily because of the ability of ionized calcium to reverse their electrical and cardiodepressant effects (1, 2). Recently, several investigators have criticized use of the term "calcium antagonist" in favor of more precise alternatives including, "slow channel inhibitor", and "calcium channel blocking drugs"(3, 4, 5). The acceptability of each of these expressions has been questioned, however, because of the absence of a proven agonist-receptor relationship and uncertainty surrounding the principle mechanisms or location whereby "calcium antagonists" interfere with calcium-dependent functions (6, 7). In this review and throughout the manuscript, the term "calcium inhibitory compound" will be used in lieu of "calcium antagonist," "slow channel inhibitor," and "calcium channel blocking drug." An exhaustive discussion of the many compounds exerting calcium inhibitory activity, their effects upon cardiovascular function, their mechanisms and sights of action, and their structure-activity relationships are beyond the scope of this manuscript. For the sake of clarity, however, a brief discussion of the role and importance of calcium in excitation-contraction coupling in cardiac tissues and several proposed mechanisms of action of calcium inhibitory compounds will be reviewed.

Ca^{2+} and Excitation-Contraction Coupling

Calcium is essential for the excitation-contraction coupling process to occur in cardiac muscle. In comparison to skeletal muscle, where intracellular calcium stores are extensive, myocardial stores of calcium are limited. Two arguments are used to stress the importance of transmembrane Ca^{2+} influx in the excitation-contraction coupling process. These are: 1) reduction in the force of cardiac contraction following the removal of Ca^{2+} from the perfusing solution, and 2) uncoupling of excitation-contraction in heart muscle by lanthanum, (a substance which displaces rapidly exchangeable membrane bound calcium, blocks calcium influx, but does not itself enter the cell) (8, 10). Transmembrane calcium influx begins during rapid depolarization (phase 0) of cardiac cells and continues for several hundred milliseconds thereafter (phase 2). The transmembrane Ca^{2+} influx, which occurs during the cardiac action potential is too small to provide sufficient Ca^{2+} to initiate contraction, but may contribute to the excitation-contraction coupling process in three ways: 1) by maintaining intracellular stores of calcium in the sarcoplasmic reticulum; 2) by inducing ("triggering") the release of Ca^{2+} from the intracellular sarcoplasmic reticulum; and 3) by modulating the release of Ca^{2+} from the sarcoplasmic reticulum (11, 13). Support for the last two of these hypotheses is based upon studies in cardiac fibers from which the sarcolemma has been removed ("skinned" fibers) in order to eliminate membrane effects, (14, 15) and modulation of cardiac action potential plateau and contraction by displacement of surface bound calcium with tri- or divalent cations having ionic radii similar to that of Ca^{2+} (e.g., La^{3+}, Co^{2+}, Mn^{2+}, Ni^{2+}) (16, 17). Studies using "skinned" cardiac fibers indicate that concentrations of Ca^{2+} substantially less than required for contraction are capable of triggering the release of Ca^{2+} from the intracellular sarcotubular system (calcium induced calcium release) (11, 15, 18). When the Ca^{2+} concentration within myocardial cells increases from resting values of 10^{-7} M to 10^{-5} M, Ca^{2+} binds to troponin releasing troponins inhibitory effects upon myosin allowing muscular contraction to occur (19) (Figure 1).

Current belief contends that the development of contractile force in cardiac muscle is determined by the amount of nonenergy dependent sarcolemmal Ca^{2+} binding. Extracellular Ca^{2+} can bind to sarcolemmal phospholipids or an external layer of material composed of glycoprotein, glycolipid and mucopolysaccharide termed the "glycocalyx." Quantitatively, the binding of Ca^{2+} to the glycocalyx is of secondary importance compared to that bound by phospholipid elements. The glycocalyx does play a significant role in the determination of myocardial cell Ca^{2+} permeability (20, 21). Upon arrival of the appropriate electrical stimulus (action potential), Ca^{2+} crosses the sarcolemma and is the principal cation responsible for a current called the "slow inward current" (I_{si}) (12, 22, 23, 24). Calcium is conducted across the sarcolemma through channels or pores which are controlled by the phosphorylation of sarcolemmal and sarcotubular proteins. Cardiac sarcolemma and sarcoplasmic reticulum are phosphorylated by exogenous and endogenous cyclic adenosine 3'-5'- monophosphate (cAMP)-dependent protein kinases (25, 26). Recently, another regulatory mechanism has been identified in cardiac sarcoplasmic reticulum. Cardiac contraction was shown to be dependent upon a multifunctional regulator protein called calmodulin (27, 28). Calmodulin-dependent phosphorylation of sarcoplasmic reticulum is dependent upon the presence of free intracellular Ca^{2+} (29, 30). Both cAMP-dependent and calmodulin-Ca^{2+} dependent phosphorylation of a proteolipid (phospholamban) of the protein kinases, result in stimulation of Ca^{2+} transport and Ca^{2+} dependent ATPase activities of sarcoplasmic reticulum (30, 31,

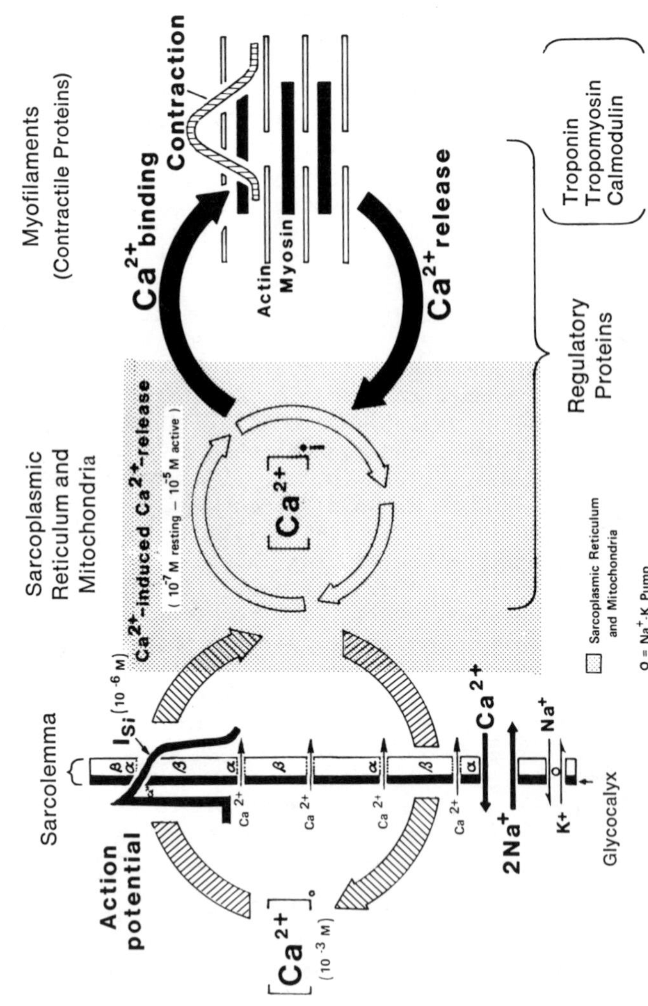

Figure 1. Illustration of the importance of Ca^{2+} in excitation-contraction coupling. See text for discussion.

32). The small amount of Ca^{2+} released from the sarcolemmal-glycocalyx complex, therefore, serves as a trigger for calmodulin phosphorylation of cardiac sarcoplasmic reticulum resulting in Ca^{2+} release from the sarcolemmal-sarcotubular system and cardiac contraction.

A second mechanism separate from the slow channel, whereby Ca^{2+} can enter the cardiac cell, is Na^{+}-Ca^{2+} exchange (33). Movement of Ca^{2+} across the sarcolemma occurs via a Na^{+}-Ca^{2+} carrier system which is dependent upon intracellular Na^{+} concentration (34). Interventions which produce an increase in intracellular Na^{+} concentration (e.g., digitalis) stimulate the carrier to move Na^{+} out and Ca^{2+} into the cell. The net inward movement of Ca^{2+} results in an increase in the force of cardiac contraction. The relative magnitude and functional importance of the Na^{+}-Ca^{2+} exchange mechanism are unclear at this time.

Proposed Mechanisms and Sight of Action of Calcium Inhibitory Compounds in Cardiac Muscle

It is currently accepted that calcium can enter the myocardial cell by three primary mechanisms including passive diffusion (extracellular $[Ca^{2+}]=10^{-3}$ M, resting intracellular $[Ca^{2+}]=10^{-7}$ M), electrogenic Na^{+}-Ca^{2+} exchange, and through voltage activated time-dependent and independent ion selective channels or pores (35). Voltage-dependent transfer of calcium ion into the cardiac cell is controlled by "gating mechanisms" and by phosphorylation of membrane proteins (30, 35). Alpha adrenergic, histaminergic, and serotonergic receptors regulate the "gating mechanisms" responsible for Ca^{2+} flux through the sarcolemnal channels (25, 36). As previously stated, transscarcolemmal Ca^{2+} may participate in myocardial excitation-contraction coupling and intracellular calcium regulation by 1) initiating the electrical event responsible for contraction; 2) replenishing intracellular calcium stores in the sarcoplasmic reticulum; and 3) modulating the release of calcium from intracellular sights.

Until recently, chemical compounds that antagnonized Ca^{2+} were believed to act by reversibly sealing specific Ca^{2+} channels at the membrane surface (sarcolemma-glycocalyx complex) of myocardial cells[1]. Divalent and trivalent cations with radii larger than the calcium ion including Mn^{2+}, Ni^{2+}, Co^{2+}, and La^{3+} will block the influx of Ca^{2+} into myocardial cells by inhibiting the passage of Ca^{2+} through membrane pores or channels (16, 36-39). The list of compounds purported to inhibit Ca^{2+} influx includes a large number of structurally related and unrelated chemical compounds including papaverine, (40) diazoxide, (41) perhexiline, (42) lidoflazine, (43) prenylamine, (4) fendiline, (4) verapamil, (38) methoxyverapamil (D-600), (38) diltiazem, (44) nifedipine, (45) niludipine,(46) nisoldipine, (47) propafenon, (48) tiapaml,

(49) ryanodine, (50) AHR-2666, (51) 2-substituted 3-dimethylamino-5,6-methylenedioxyindenes (MDI) (52, 53) and many others (Figure 2) (53, 54). Recently, this relatively simplistic classification of calcium inhibition has been challenged by studies which indicate that many of the so-called "calcium antagonists" may act intracellularly (38, 55-58). Several arguments are used to support the hypothesis of an intracellular mechanism of action including: 1) at 4 mM extracellular potassium concentrations excitation-contraction coupling is predominantly dependent upon intracellular calcium stores (59); 2) most calcium inhibitory compounds demonstrate inhibition of the contractile response (electrical-mechanical uncoupling) at concentrations which do not affect the plateau phase of the cardiac action potential (1, 60, 61, 62) 3); studies using mesenteric vascular smooth muscle indicate that low concentrations of calcium inhibitory compounds interfere with phasic contractions which are principally dependent upon intracellular calcium activity, while high concentrations of calcium inhibitory compounds interfere with tonic mechanical activity which is dependent upon extracellular calcium (58); 4) radiolabeled calcium (^{45}Ca) transmembrane flux is not inhibited by calcium inhibitory compounds (3); 5) 2-n-propyl and 2-n-butyl MDI bind to calmodulin and troponin (M.T. Piascik *et al.*, manuscript in preparation); 6) Lineweaver-Burk plots demonstrate that verapamil and D-600 do not compete with Ca^{2+} for the same membrane site and that the onset of drug effect is much slower than that of Mn^{2+} (38); and 7) the observation that the inhibition of transsarcolemmal Ca^{2+} flux by verapamil is a direct function of tissue preparation pacing frequency independent of its direct negative inotropic activity (63). The precise mechanism and sight of action of most compounds categorized as calcium inhibitory compounds, therefore, remains obscure. Future refinements in experimental models and techniques will undoubtedly will lead to the classification of calcium inhibitory compounds based upon their primary mechanism of action and specific site(s) of action (extracellular *vs.* intracellar). Because of the uncertainty surrounding the precise mechanisms of action of calcium inhibitory compounds, I will describe their cardiac electrical and mechanical effects illuding when possible to those compounds that are believed to act: 1) competitively with Ca^{2+} for specific calcium channels (e.g., Co^{2+}, Mn^{2+}, La^{3+}, etc.); 2) at the cardiac cell membrane and possibly by one of several intracellular mechanisms (e.g., verapamil, diltiazem, nifedipine); and 3) intracellularly (e.g., 2-n-propyl and 2-n-butyl MDI).

Normal and Abnormal Cardiac Electrical Activity

Most studies attempting to define the effects of calcium inhibitory compounds upon cardiac electrical activity have focused upon the plateau phase of the cardiac action potential. In cardiac and other excitable tissues, the action potential represents multiple ionic transmembrane fluxes (35). Microelectrode and voltage clamp experiments have confirmed that two different inward currents are responsible for the development of the cardiac action potential (Figure 3) (12, 24, 35, 36, 37, 64). A rapid inward current carried by Na^+ (I_{na^+}) is responsible for the initial depolarization or spike of the cardiac action potential. The transfer of Ca^{2+} across the sarcolemma gives rise to a second inward current that with a small Na^+ component is referred to as the "slow inward current" (I_{si}). The adjective, slow, is used because of the slow kinetics of calcium ion transfer compared to the rapid Na^+ influx which occurs during the initial depolarization. For descriptive purposes, the cardiac action potential is subdivided into 5 distinct phases. During phase 0, transmembrane conductance for the inward displacement of Na^+ rapidly increases (rapid inward current I_{Na^+}) resulting in rapid depolarization. As Na^+ reaches its equilibrium potential (+20-40 mV), as predicted by the Nernst Equation, Na conductance decreases. Rapid inactivation of Na^+ conductance coupled with the cellular influx of chloride ion results in a brief period of rapid repolarization, referred to as phase 1. In Purkinje fibers, rapid sodium inactivation is delayed resulting in a very long-lasting depolarizing or inward current. During phase 0 when membrane potential has depolarized to a value of from -60 to -40 mV, the second slower inward current (I_{si}) is activated. I_{si} is responsible for the prolongation of membrane depolarization or plateau phase (phase 2) of the cardiac action potential. Experimentally, divalent ions with similar radii to that of Ca^{2+}, such as strontium (Sr^{2+}) and barium (Ba^{2+}), have been used as charge carriers for I_{si}. Repolarization occurs when the sum of the outward currents exceeds that of the inward currents (36). At membrane potentials less negative than -40 mV, transmembrane conductance to K^+ increases. Increased K^+ conductance and the resultant outward displacement of K^+ coupled with a simultaneous decrease in I_{si} (calcium inactivation) lead to cellular repolarization (phase 3). Transmembrane conductance to K^+ and intracellular calcium activity are significant factors in determining the rate of repolarization (65-68). For example, experimental or pharmacologic manipulations which increase intracellular calcium activity or increase membrane conductance to K^+ markedly shorten action potential duration (69, 70). The remaining phase of the cardiac action potential is referred to as phase 4. In resting atrial and ventricular myocardium, phase 4 is stable at a resting membrane potential ranging from between -80 to -95 mV. In cardiac tissues demonstrating spontaneous activity (i.e., sinoatrial [SA] node, atrioventricular [AV] node, specialized conducting [Purkinje]

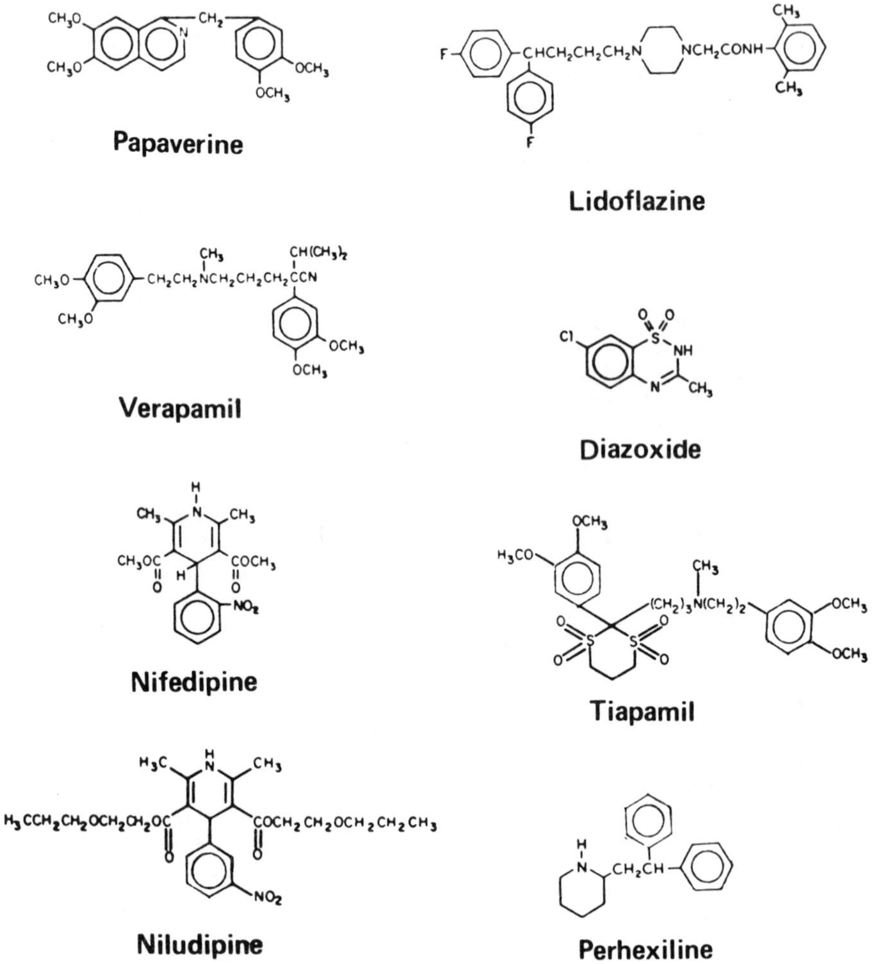

Papaverine

Lidoflazine

Verapamil

Diazoxide

Nifedipine

Tiapamil

Niludipine

Perhexiline

Figure 2. Structural formulas of some calcium inhibitory compounds. Continued on next page.

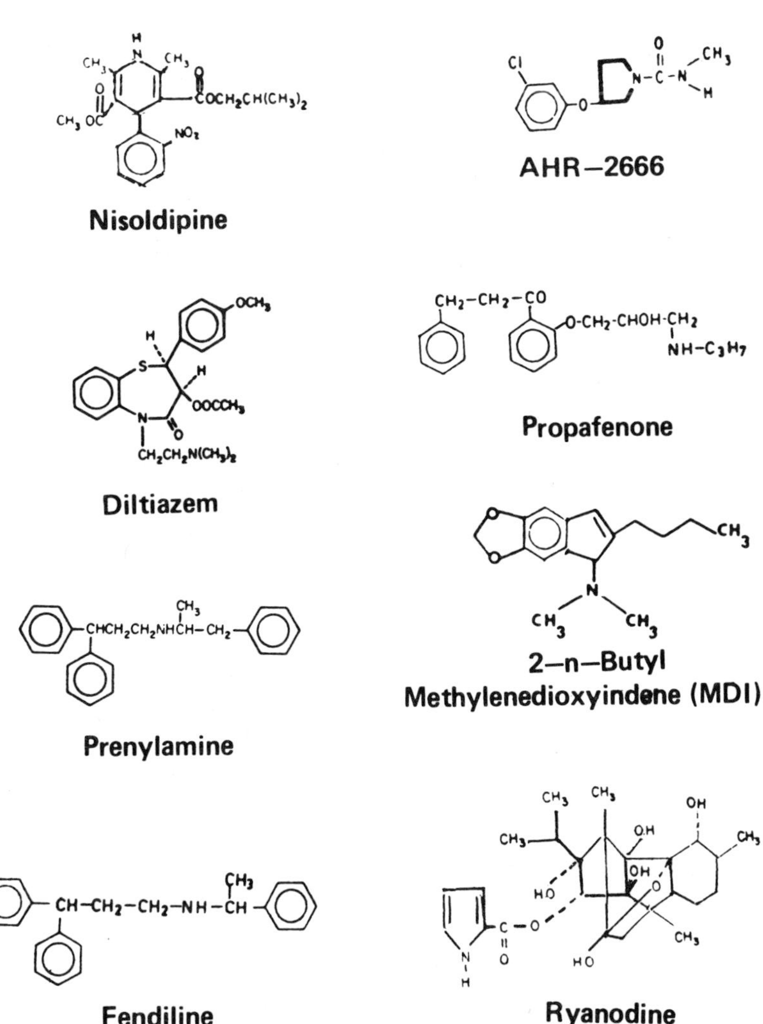

Figure 2. Continued. *Structural formulas of some calcium inhibitory compounds.*

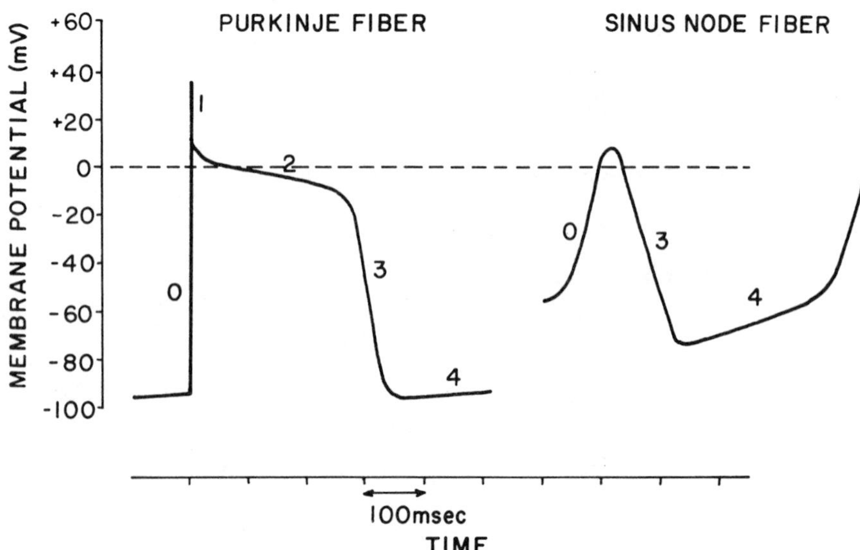

Figure 3. Action potentials recorded from Purkinje and SA nodal fibers. Note the absence of phase 1 and slurring together of phases 2 and 3 in the recording from the SA nodal fiber.

fibers), phase 4 is characterized by a gradual change in the membrane potential towards a less negative or threshold potential from which an action potential is generated. Phase 4 activity, called diastolic depolarization, is responsible for the so-called "pacemaker current" of cardiac cells, and in Purkinje fibers has been attributed to a decaying outward K^+ current, termed I_{K2} (71). Recently an entirely new mechanism for the pacemaker current and phase 4 activity in Purkinje fibers has been proposed (72, 73). Phase 4 is attributed to a mixed current principally generated by activation of an inward Na^+ component rather than inactivation of I_{K2} (73). Acceptance of this new concept has clarrified several important qualitative differences thought to exist between phase 4 activity in cells of the SA node and His-Purkinje system.

The shape of the normal cardiac action potential varies considerably depending upon the recording sight. Cells of the SA and AV nodes demonstrate a maximum diastolic potential of -60 mV compared to -90 mV recorded from Purkinje and muscle fibers, a markedly reduced rate of rapid depolarization (phase 0), no phase 1 and a slurring together of phases 2 and 3. The characteristic, reduced, resting membrane potential and slow rate of phase 0 in these tissues have led to the conclusion that phase 4 is partially dependent upon the inward displacement of calcium ion and that phase 0 is due to the inward movement of Na^+ through partially inactivated fast channels and Na^+ and Ca^{2+} through slow channels (74).

Alterations in normal impulse formation are due to normal and abnormal automaticity, altered conduction, or both, and are responsible for abnormal electrical activity and the many diverse arrhythmias that occur in diseased hearts (75). In addition, there is suggestive evidence that cardiac arrhythmias result from electrical uncoupling of the ventricular myocardium (76, 77). Electrical uncoupling of cardiac cells occurs simultaneously with increases in internal cellular resistance and Ca^{2+} concentrations (77, 78). Interventions that exaggerate increases in internal resistance including hypoxia, ischemia, catecholamines, increases in extracellular Ca^{2+} and increased stimulation frequency also increase intracellular Ca^{2+} (75). The cellular electrophysiologic changes that are believed to have an important role in the development of abnormal automaticity and conduction, and the development of cardiac arrhythmias in diseased hearts include variable reductions in resting membrane potential, altered action potential configuration and duration, unprovoked and triggered spontaneous activity, depression of cellular excitability and membrane responsiveness, prolongation of action potential refractoriness and action potentials with a markedly reduced rate of rise of phase 0 (75). Acute cellular damage from any cause is generally associated with a loss of cell membrane integrity and a decrease in intracellular K^+ concentration.

These changes, combined with local increases in extracellular K^+ concentration, result in cellular depolarization, a decreased rate of rise of phase 0, decreased action potential duration and potentially the development of calcium dependent (slow response) action potentials (75). In more chronic (hrs or days) situations, after extracellular K^+ concentration has returned to normal, transmembrane potential may be normal or reduced, but action potential duration may be prolonged due to decreases in the transmembrane K^+ concentration gradient and reductions in membrane conductance to K^+ (75). In either case, a reduction in resting membrane potential results in a cardiac action potential more dependent upon Ca^{2+} for depolarization and predisposes the cardiac cell to spontaneously occurring cyclical membrane oscillations (depolarization induced automaticity) which may reach threshold and produce extradepolarizations (75, 79, 80). Cardiac cells that develop calcium dependent action potentials may also demonstrate spontaneous or triggered (dependent on a prior initiating action potential) abnormal automaticity, delayed conduction patterns, and prolonged refractoriness. All of these factors are important in the development of cardiac arrhythmias (80-83). Independent of disease induced effects, pharmacologic interventions (digitalis, barium, catecholamines) can potentiate the slow inward Ca^{2+} current in otherwise normal cardiac tissues and may trigger membrane oscillations which lead to abnormal cellular automaticity and cardiac arrhythmias (81).

Drugs that demonstrate calcium inhibitory activity display different cardiac membrane effects dependent upon their structure, physical properties and concentration. Compounds which interfere with Ca^{2+} influx are expected to prolong the duration of the action potential. This prediction is supported by evidence that decreased intracellular Ca^{2+} activity may decrease the late increase in K^+ conductance and potentially prolong the duration of repolarization (phase 3) (69). The first substances reported to inhibit I_{si} without influencing rapid depolarization (phase 0) were Mn^{2+}, Ni^{2+} and Co^{2+} (16, 17). The effects of the chemically complex calcium inhibitory compounds upon I_{si} and action potential duration, however, are difficult to predict because they may not act by a single electrophysiologic mechanism. Finally, it has been suggested that the Na^+-Ca^{2+} exchange mechanism may be important in providing calcium for cardiac contraction (33). Metallic cations (Mn^{2+}, $N_i{}^{2+}$, Co^{2+}) are known to displace membrane bound Ca^{2+} which participates in Na^+-Ca^{2+} exchange (34, 39). The effect of the majority of calcium inhibitory compounds on the Na^+-Ca^{2+} exchange process is uncertain.

Electrophysiologic Effects of Calcium Inhibitory Agents in Normal and Diseased Hearts

Effects Upon Action Potential Configuration and Automaticity. The one electrophysiologic effect that all calcium inhibitory compounds apparently share in common is inhibition of I_{si}. The magnitude of this effect is dependent upon the potency and dose of the drug being investigated. Furthermore, a number of compounds with calcium inhibitory activity demonstrate multiple membrane effects dependent upon optical isomerism. The (+)-isomers of racemic verapmil, D-600, and prenylamine demonstrate predominantly fast-channel (I_{Na^+}) inhibitory effects whereas, the (-)-isomers of these same compounds are predominantly inhibitors of I_{si} (43, 84, 85). Interestingly, only those calcium inhibitory compounds which demonstrate effects upon I_{Na} are useful clinically as antiarrhythmics (43, 86-89). Because of differences in drug potency, dose, and formulation, descriptive experiments report that calcium inhibitory compounds have variable effects upon action potential characteristics and configuration (1, 60, 62, 90, 91, 92). In general, several statements can be made about compounds exhibiting calcium inhibitory activity. At drug concentrations, which begin to result in negative inotropic effects in isolated heart muscle preparations, most calcium inhibitory compounds cause little or no change in action potential amplitude, maximum upstroke velocity of phase 0 (V_{max}), or action potential duration at 100 percent repolarization. As drug concentration is increased, gradual changes in atrial, ventricular myocardium and Purkinje fiber action potential configuration are observed. In atrial and ventricular myocardium, these changes include increases in the slope of the plateau (phase 2), a decrease in action potential duration at 25 and 50 percent repolarization (ADP_{25} and APD_{50}) and decreases, no change, or increases in APD_{100}. At still higher drug concentrations, the demarcation between phases 2 and 3 becomes indistinguishable and APD is decreased at all points during repolarization. Associated with the changes observed in APD_{100} are similar, but not proportional, changes in the effective refractory period (ERP); decreases in APD_{100} result in increases in the ERP/APD. In addition, several compounds (2-n-propyl and 2-n-butyl MDI, racemic verapamil, D-600, prenylamine) decrease V_{max} (43, 84, 91). Similar but more pronounced changes are observed in action potentials recorded from Purkinje fibers.

Although studies using voltage clamp techniques suggest that several of the calcium inhibitory compounds (verapamil, D-600, and manganese) may affect background inward currents carried by Na^+ or time-dependent outward currents carried by K^+ (both of which could affect action potential duration), the changes that occur in action potential configuration could be caused by variations in the degree of intracellular Ca^{2+} inhibition (93, 94). Drugs which have profound depressant effects upon intracellular Ca^{2+} activity would be expected to dramatically abbreviate phase 2 leading to premature activation of the outward K^+ currents and marked reductions in APD during all phases of repolarization. Compounds which have poor intracellular Ca^{2+} inhibitory activity and which act primarily at the cell membrane by modulating the influx of Ca^{2+}, indirectly reduce intracellular Ca^{2+} release from sarcoplasmic reticulum. This results in minimal abbreviation of phase 2, reductions in APD_{25} and APD_{50}, and delayed activation of outward K^+ currents, thereby prolonging APD_{100}. Recently, we have had the opportunity to examine the *in vitro* electrophysiologic effects of two calcium antagonists which are believed to act intracellularly (2-n-propyl and 2-n-butyl MDI) (55). Addition of either of these compounds at concentrations ranging from 10^{-7} to 10^{-5} M to Tyrode's solution superfusing canine Purkinje fibers results in action potential changes characteristic of calcium inhibition, including decreases in APD_{25}, APD_{50} and APD_{100}. In separate experiments, similar concentrations of a quaternary derivative of 2-n-butyl MDI were added to the superfusion medium resulting in significant reductions in APD_{25}, variable changes in APD_{50} and increases in APD_{100} (Figure 4). These experiments suggest that the electrophysiologic changes caused by calcium inhibitory compounds are in part dependent upon inhibition of I_{si} and the currents responsible for repolarization. Because of the negligible effects of 2-n-butyl MDI and its quaterinary derivative upon phase 4 depolarization (discussed below), it is unlikely that it interferes with action potential repolarization by altering K^+ efflux, although this aspect requires further study.

Effect Upon Normal Automaticity

The effects of many Ca^{2+} inhibitory compounds on sinoatrial, atrioventricular, and Purkinje fiber automaticity have been examined (43, 46, 95-100). In general, *in vitro* studies suggest that the addition of Ca^{2+} inhibitory compounds to solutions superfusing isolated tissues results in no change or a decrease in spontaneous activity (98, 100). As previously stated, time-dependent outward currents carried by K^+, and activation of an inward current carried by Na^+ and Ca^{2+} exert a significant influence on SA and AV node spontaneous activity (74). Tyrodes superfused in rabbit and guinea pig and blood perfused canine sinus node preparations display the most pronounced negative chronotropic effects to compounds which have the ability to inhibit Na^+ and Ca^{2+} transmembrane flux. Verapamil, D-600, diltiazem and niludipine, in addition to inhibiting of I_{si} induce changes in Na^+ and, indirectly in K^+ transmembrane flux resulting in pronounced decreases in spontaneous rate (46, 91, 102). In contrast to their effects upon SA and AV automaticity, calcium inhibitory compounds do

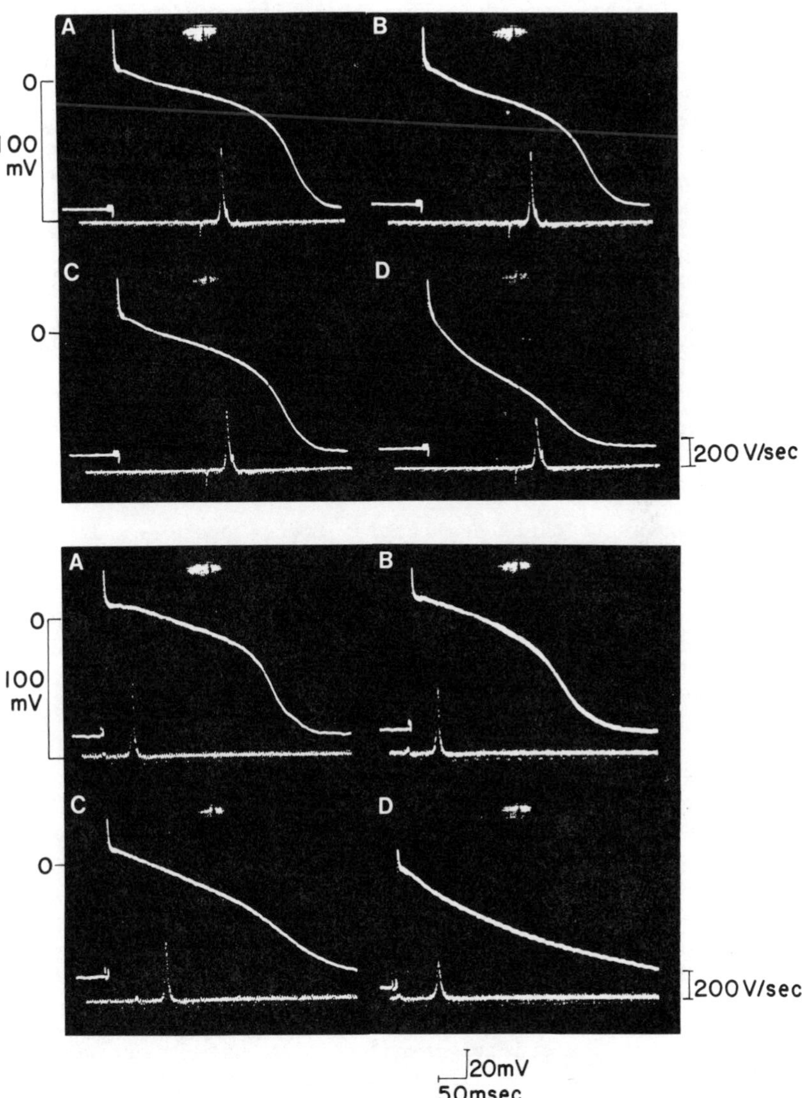

Figure 4. Action potentials recorded from Purkinje fibers before (A) and after 1×10^{-6} M *(B), 1×10^{-5}* M *(C), and 1×10^{-4}* M *(D) 2-*n*-butyl MDI (top) and quaternary 2-*n*-butyl MDI (bottom). Note the obvious prolongation in APD_{100} in C(bottom).*

not affect, or only minimally suppress, spontaneous activity in normal Tyrodes or blood superfused Purkinje fibers (51, 60, 61, 62, 102). This finding is not suprising in view of the lack of importance of the inward Ca^{2+} current in determining spontaneous activity in normal Purkinje fibers. Those calcium inhibitory compounds that do exhibit the ability to depress spontaneous normal automaticity in Purkinje fibers also exhibit effects upon membrane Na^{+} conductance. Nifedipine, a compound believed to act primarily by inhibiting Ca^{2+} influx, has no effect upon Purkinje fiber automaticity (60). Verapamil, a calcium inhibitory compound with both Na^{+} and Ca^{2+} inhibitory activity, has been reported to cause no change or decreases in Purkinje fiber automaticity (62). Both 2-n-propyl and 2-n-butyl MDI, and AHR-2666 demonstrate insignificant effects upon the slope of phase 4 but depress automaticity in Purkinje fibers. Calcium antagonists that depress spontaneous activity without affecting phase 4 depolarization presumably act by shifting the threshold for activation of the inward Na^{+} current in a positive direction (51).

Effects Upon Abnormal Automaticity

In contrast to their negligible effects upon normal automatic mechanisms in Purkinje fibers, calcium inhibitory compounds depress abnormal automaticity induced by a variety of experimental techniques (51, 60, 61, 62, 99-102). Calcium inhibitory compounds are particularly useful in suppressing impulse formation due to intrinsic changes in membrane conductance which result in cyclical, pacemaker-like oscillations of diastolic potential (51, 80, 91). This type of activity is most commonly observed in depressed, partially depolarized fibers and has been termed "depolarization-induced automaticity" (103). Depending upon their temporal relationship to normally produced action potentials, membrane oscillations are called early or delayed afterdepolarizations (104). Delayed afterdepolarizations are also called oscillatory afterpotentials, transient depolarizations, low amplitude potentials and enhanced diastolic depolarizations (104). Afterdepolarizations that are dependent on a prior initiating action potential, reach threshold and initiate rhythmic activity are called "triggered" (83, 105). Afterdepolarizations and triggered activity have been recorded from SA and AV nodes, specialized atrial and ventricular (Purkinje) fibers, atrioventricular valve leaflets, coronary sinus tissue, and ventricular myocardium (51, 75, 80, 105, 106). In a recent review, it was pointed out that membrane oscillations can be produced by stretch, ischemia, hypoxia, cooling, drugs (catecholamines, digitalis, Ca^{2+}, Ba^{2+}, aconitine, veratrine), elevated P_{CO_2}, alterations in the electrolyte content of superfusion solutions, and spontaneously occurring disease (75). The calcium inhibitory compounds which have demonstrated the ability to reduce or eliminate membrane oscillations include metallic ions (Mn^{2+}, Ni^{2+}, Co^{2+}), verapamil, D-600, nifedipine, niludipine, diltiazem, perhexiline, AHR 2666, 2-n-propyl and 2-n-butyl MDI and many others (51, 60, 61, 62, 99-108). Recent evidence suggests that transient release of intracellular Ca^{2+} from sarcoplasmic reticulum is responsible for the membrane conductance changes which produce membrane oscillations (104). If this theory is correct, the efficacy of calcium inhibitory compounds in reducing and abolishing membrane oscillations can be linked to their ability to limit intracellular calcium release, transmembrane Ca^{2+} influx, or both. Since partial activation of the background inward Na^{+} current is believed to be somewhat responsible for membrane oscillations, calcium inhibitory compounds which inhibit Na^{+} transmembrane flux (verapamil, D-600, perhexiline, AHR 2666, 2-n-propyl and 2-n-butyl MDI's) may be particularly useful in abolishing abnormal rhythm disturbances due to membrane oscillations (79).

Effects Upon Conduction of the Cardiac Impulse in Normal and Diseased Hearts

Cardiac activation is dependent upon the proper timing and sequencing of cardiac excitation. Alterations in the pathways or conduction velocity which result in normal cardiac activation could lead to abnormal conduction patterns and disturbances in cardiac rhythm (75). The relative potencies of calcium inhibitory compounds in depressing cardiac conduction velocity in isolated tissues is dependent upon dose and the type of tissue being investigated. Normal atrial and ventricular muscle, and Purkinje fibers demonstrate little or no response to the negative dromotropic effects of calcium inhibitory agents unless exposed to extremely large drug concentrations (62, 100, 102, 109). Tyrode's or blood superfused SA and AV nodal tissues demonstrate dose dependent decreases in conduction velocity when exposed to calcium inhibitory compounds (96-100, 102). Recordings from the upper and middle portions of the AV node indicate that calcium inhibitory compounds cause a reduction in action potential amplitude and upstroke velocity and increase the effective refractory period (102). The ionic mechanisms responsible for the effects of calcium inhibitory compounds upon conduction velocity in SA and AV nodal tissues remain undetermined although inhibition of I_{si} is conjectured to be responsible for the majority of changes based upon reversal of these effects by increasing extracellular Ca^{2+} or catecholamine administration (46, 102, 104).

The electrophysiologic correlates of the effects of calcium inhibitory agents upon conduction velocity in isolated cardiac tissues are observed in isolated whole heart preparations (electrograms), and specific intervals of the electrocardiogram recorded from intact animals (48, 95, 96, 98, 99, 110-113). Drug effects are dependent upon diverse hemodynamic, metabolic, autonomic and reflex changes. For example, *in vitro* studies indicate that nifedipine is twice as potent as verapamil in slowing AV nodal conduction when both are administered at equal negative chronotropic concentrations (95). *In vivo* studies, on the other hand, indicate that nifedipine increases heart rate and does not affect AV nodal conduction, while verapamil prolongs AV nodal conduction and produces second degree AV block (89, 111, 113). Furthermore, verapamil exerts its negative dromotropic effects in both innervated and denervated hearts (111). Studies in patients in sinus rhythm support these findings by indicating that while verapamil has no effect upon the R-R, QRS and Q-T_c intervals of the electrocardiogram, it dramatically prolongs both the atrial to His (A-H) time and the P-R interval (39, 111). Similar studies using nifedipine in resting human patients are unable to demonstrate discernable effects upon SA and AV nodal or His Purkinje conduction (96). The calcium inhibitory compounds, diltiazem and perhexiline, demonstrate effects in intact hearts midway between those of verapamil and nifedipine (6, 98). Several investigators attribute the differences caused by calcium inhibitory compounds in intact animals to the magnitude of their *in vitro* fast (Na^+) versus slow channel (Ca^{2+}) inhibitory activity and their ability to initiate reflex autonomic effects (114, 115). For example, verapamil inhibits both I_{Na} and I_{si} *in vitro*, but demonstrates minimal autonomic effects *in vivo* (62, 111). Nifedipine, however, depresses I_{si} *in vitro* and elicits strong reflex sympathetic activity *in vivo* (114). Studies of these calcium inhibitory compounds which are believed to act intracellularly (2-n-propyl and 2-n-butyl MDI's) upon automaticity and conduction in SA and AV nodes and ischemic myocardium remain to be done.

The effects of calcium inhibitory agents upon conduction velocity and delayed activation in normal hearts may be entirely different from that observed in ischemic myocardium. In constrast to depressant I_{si} dependent conduction effects in SA and AV nodes of normal hearts, verapamil has demonstrated the ability to reduce the degree of ischemia-induced conduction delay in anesthetized dogs after ligation of their left anterior descending coronary artery (109, 110). Recent studies using nifedipine, diltiazem, nisoldipine and niludipine have yielded results suggesting a similar effect upon ischemia-induced conduction delay (109, 116-124). These results imply that calcium inhibitory compounds favorably affect myocardial oxygen supply and demand presumably by improving oxygen supply to ischemic areas via coronary collateral blood flow. Furthermore, it has been argued that the administration of calcium inhibitory compounds prior to infarction improves recovery of myocardial function after coronary artery ligation by decreasing intracellular Ca^{2+} and energy demand during the ischemic period (78, 122, 125). The relative potency of the various calcium inhibitory compounds in producing this myocardial "sparing" or "salvaging" effect has not been determined.

Antiarrhythmic Activity

Calcium inhibitory compounds are used as antiarrhythmic drugs because of their diverse effects upon cardiac electrical activity and experimental evidence which suggests that the slow inward current is important in the genesis of cardiac arrhythmias (126). A variety of potentially toxic compounds including aconitine, barium, digitalis and catecholamines have been used to produce abnormal electrical activity in isolated tissue preparations in order to demonstrate the effectiveness of calcium inhibitory compounds in suppressing abnormal automaticity. Experimental studies in intact animals indicate that calcium inhibitory compounds possess variable anti-arrhythmic effects against a variety of chemically- (digitalis, calcium, catecholimines) and mechanically- (coronary occlusion) induced ventricular arrhythmias (79, 81, 109, 110, 127, 128, 129.). The importance of I_{si} in the genesis of ventricular arrhythmias associated with these pharmacologic and mechanical manipulations is assumed but remains speculative. The clinical value of the various calcium inhibitory compounds in effectively controlling cardiac arrhythmias caused by toxic concentrations of digitalis and naturally-occurring disease (ischemia, infection, hypoxia) is also disputed (1-26, 130). It is noteworthy that calcium inhibitory compounds with the ability to inhibit I_{Na^+} possess the most potent antiarrhythmic activity against both experimental and clinical arrhythmias (126). For example, nifedipine exerts minimal, if any, antiarrhythmic effect against ischemic induced arrhythmias in intact animals (6, 87, 131). Diltiazem and perhexiline, on the other hand, demonstrate variable antiarrhythmic effects dependent upon their dose and the mechanism responsible for arrhythmia production (ischemia, hypoxia, digitalis, aconitine) (102, 126, 131, 132). Only verapamil, which has been investigated in the largest number of experimental and clinical trials, has demonstrated consistent antiarrhythmic activity against cardiac arrhythmias regardless of cause (133). Clinical electrophysiologic studies indicate that verapamil is effective in combating supraventricular arrythmias including atrial premature depolarizations, paroxysmal atrial tachycardia, AV nodal re-entrant paroxysmal supraventricular tachycardia, atrial fibrillation and atrial flutter (134). Verapamil is also potentially beneficial in eliminating recurrent supraventricular tachycardia in patients with accessory pathways and pre-excitation (135). Verapamil is most effective in controlling circus movement tachycardia involving accessory pathways when retrograde conduction is involved rather than when conduction is antegrade. Verapamil will not slow ventricular rate during atrial fibrillation or antegrade conduction over the accessory pathway. In addition, since large doses of verapamil may shorten the cardiac action potential and the effective refractory period, increases in ventricular rate have been reported (134, 135). Recent experimental evidence using 2-n-propyl and 2-n-butyl MDI indicates that these compounds possess antiarrhythmic potency equivalent to that of verapamil without producing bradycardia, ECG changes or atriovenricular block (52, 136). It will be interesting to determine if the MDI's are equally as effective in combating arrhythmias in patients with naturally occurring heart disease as they are in experimental animals.

In summary, compared to diltiazem, nifedipine, and perhexiline, verapamil demonstrates the most consistent results in the control of both experimental and clinical arrhythmias. In none of the studies reported to date, however, has verapamil been uniformly successful in re-establishing normal sinus rhythm in all animals or patients although there is a marked reduction in the frequency of ectopic ventricular depolarizations (126, 134). The important direct antiarrhythmic effects of the various calcium inhibitory agents are the result of a combined inhibition of both I_{Na^+} and I_{si} (126). Additional explanations for the direct antiarrhythmic effects of calcium inhibitory agents include hyperpolarization of the cardiac cell membrane thereby reducing calcium cellular influx and membrane oscillations (57, 60, 62, 110), increases in calcium uptake or decreased calcium release from sarcoplasmic reticulum or mitochondria (7, 78, 100), stimulation of the sodium-calcium pump promoting calcium efflux (33), interference with the intracellular calcium receptor of contractile proteins, and binding to intracellular regulatory proteins (M.T.Piascik *et al*, manuscript in preparation). Equally as important, and potentially more clinically relevant, are the indirect mechanisms responsible for the antiarrhythmic activity of calcium inhibitory agents, including coronary vasodilation and subsequent increases in coronary artery blood flow (47, 114, 124, 137); decreased cardiac contractillty, metabolism, and arterial vasodilation resulting in decreases in myocardial oxygen consumption and wall tension; (92) and reflex alterations in autonomic regulation of heart rate (overdrive supression) and AV conduction (114).

Cardiac and Smooth Muscle Mechanical Activity

Cardiac Muscle. Calcium inhibitory agents may interfere with excitation-contraction coupling processes in myocardial or vascular smooth muscle cells by a number of mechanisms including: 1) inhibition of the slow inward current through direct competition for slow channels or interference with the membrane binding of Ca^{2+}; 2) interference with the release and uptake of calcium by the various intracellular organelles; and 3) alteration of the activity of specific regulatory proteins (troponin, calmodulin). It is interesting to note that most calcium inhibitory agents are from 3 to 10 times more effective in reducing smooth muscle contraction than in reducing the force of myocardial contraction. Furthermore, unlike cardiac muscle where tropomyosin and troponin are believed to be the important regulatory proteins, a substance called calmodulin serves a predominant role in smooth muscle (138, 139).

Experiments utilizing isolated superfused and blood perfused cardiac tissue preparations and isolated rat, rabbit, and cat whole hearts all demonstrate the potent, concentration-dependent, negative inotropic activity of the calcium inhibitory compounds (45, 50, 90-95, 112, 113, 114, 118, 140, 141). Studies comparing the relative negative inotropic effects of several of the calcium inhibitory compounds indicate that those compounds exhibiting the most potent calcium inhibitory effects *in vitro* are the most effective in reducing cardiac contractility *in vivo* (6). Of those compounds most frequently compared, nifedipine and niludipine exhibit the most profound negative inotropic activity followed by verapamil (levo isomer), diltiazem and perhexiline (88, 89-142).

In isolated guinea pig, rat, rabbit and cat papillary muscle preparations, the addition of a calcium inhibitory compound to the superfusing solution decreases isometric or isotonic tension long before measurable changes in the cardiac action potential occur (50, 89, 92, 118, 142). The subsequent administration of catecholamines or increases in the extracellular Ca^{2+} concentration (2.5 to 5.0 mM) will partially reverse these effects and originally served as the basis for the hypothesis that calcium inhibitory compounds limit sarcolemmal Ca^{2+} flux via slow channels (1). As previously suggested, intracellular mechanisms are important in determining the contractile effects of calcium inhibitory compounds. Experiments conducted in our laboratory, studying the effects of 2-n-propyl or 2-n-butyl MDI in isolated Tryrodes superfused canine papillary muscle preparations, demonstrate their ability to uncouple excitation-contraction coupling at drug concentrations which do not reduce action potential characteristics including action potential amplitude, resting potential, duration at 25, 50, 90 percent repolarization and the rate of rise of phase 0 (dV/dt_{max}). Interestingly, at drug concentrations greater than 5 X 10^{-5} M, ventricular muscle action potential durations actually increased (Figure 5). The explanation for this latter effect is uncertain, although similar changes appear to be evident upon close inspection of figures reported by other investigators using a variety of calcium inhibitory compounds (1, 92).

It is established that Ca^{2+} and K^+ are involved in maintenance and termination of the plateau phase of the cardiac action potential. Furthermore, intracellular calcium concentration controls membrane K^+ permeability via the various conductance components for K^+ (gK_1, gK_2, gI_x) (69). It is also established that action potential duration and myocardial tension development are integrally related (13). In view of previous observations and explanations for the excitation-contraction coupling process and the effects of calcium inhibitory compounds upon the cardiac action potential of ventricular muscle and Purkinje fibers, one possible explanation for the effects observed in ventricular muscle preparations is that low concentrations of calcium inhibitory compounds reduce the amount of intracellular free calcium in the vicinity of the K^+ channel, thereby changing the channel's configuration resulting in a reduction in gK^+ and delayed repolarization of the ventricular muscle action potential. When elevated concentrations of calcium inhibitory compounds are used, the plateau phase of the action potential is dramatically abbreviated resulting in reductions in action potential duration during all phases of repolarization. Depending upon the cardiac fiber type (Purkinje fiber or ventricular muscle) and concentration of the calcium inhibitory compound under investigation, quantitative and qualitative changes in action potential plateau and repolarization are to be expected, although contractile force is invariably reduced.

Smooth Muscle. All known calcium inhibitory compounds decrease smooth muscle tone in both the coronary and peripheral circulations (113, 114, 115, 140, 143-147). Comparative studies using dogs suggest that this effect varies in intensity but surpasses negative inotropic activity (114, 145). Nifedipine, verapamil and perhexiline exhibit their most pronounced vasodilator effects upon femoral arterial blood flow followed by coronary, renal, and mesenteric beds (44, 47, 115). Diltiazem and nisoldipine preferentially dilate the coronary bed (47). The coronary smooth muscle relaxing effects of the 2-substituted aminoindenes closely resemble that produced by verapamil and prenylamine (53, 113). Two-n-butyl MDI, in particular, demonstrates a 4-fold greater coronary arterial vasodilating than negative inotropic effect (113). Equally as interesting and potentially of greater clinical significance than their ability to dilate coronary arteries, is the ability of several calcium inhibitory compounds to affect regional myocardial perfusion in such a way as to increase blood flow to ischemic myocardium (47, 148, 149, 150).

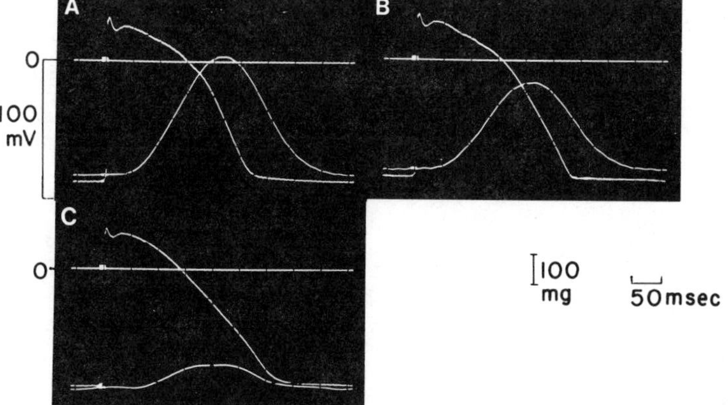

Figure 5. Action potentials and developed tension recorded from canine papillary muscle before (A) and after 1 × 10^{-6} M *(B) and 1 × 10^{-5}* M *(C) 2-*n*-butyl-MDI. Developed tension decreased by 25% (B) and 80% (C).*

Nisoldipine, nifedipine and verapamil have recently been shown to increase total coronary collateral transmural blood flow to ischemic areas of myocardium (47). Only diltiazem did not produce an increase in total transmural blood flow within the ischemic zone, although the ratio of blood flow in the subendocardium versus that in the subepicardium was significantly increased. Recent studies evaluating the effects of nifedipine in dogs with experimentally induced acute myocardial ischemia and infarction indicate that the dosage of calcium inhibitory compound utilized may be critical in determining whether or not a beneficial effect is obtained (122). Dogs receiving nifedipine, 13 μg/kg, demonstrated a 30 percent fall in aortic pressure, a 12 percent rise in heart rate, and an extension of their infarct zone. Dogs receiving nifedipine, 1 μg/kg, demonstrated a 12 percent fall in arterial blood pressure, no change in heart rate, and an improvement of regional myocardial perfusion suggesting limitation of infarct size. Together these studies indicate that drug, drug dose, dosage rate and method of administration are critical in determining whether or not a favorable redistribution of blood flow to ischemic myocardium is obtained.

Calcium Inhibitory Compounds and Myocardial Ischemia

Regardless of the effects of calcium inhibitory compounds upon total myocardial perfusion and the distribution of transmural blood flow, several groups of investigators have demonstrated that verapamil and diltiazem can prevent the consequences of myocardial ischemia, particularly mitochondrial swelling and destruction (150-153). Myocardial calcium levels in ischemic dog hearts increase to 10 to 20 times normal levels (78, 154). Increased intracellular calcium increases intracellular energy utilization, decreases ATP, and impairs mitochondrial function (78). Recent evidence suggests that ischemia of myocardial tissue results in an increase in inorganic phosphate concentration which induces mitochondrial swelling and uncoupling of oxidative phosphorylation mechanisms by stimulating energy-dissipating intramitochondrial cycling of calcium (148). Furthermore, a direct relationship exists between the accumulation of mitochondrial Ca^{2+} and the development of contracture in myocardial muscle strips and the degree of ventricular muscle stiffness in intact hearts (155). Pretreatment of tissue preparations or isolated hearts with a variety of calcium inhibitory compounds including verapamil, diltiazem, nifedipine, prenylamine, fendiline and bencyclane prevents mitochondrial swelling and uncoupling of oxidative phosphorylation and reduces intracellular calcium (155, 156, 157). The net results of these effects are an increase in adenine nucleotides, tissue ATP, creatine phosphate and improvement in cardiac function (78).

Calcium Inhibitory Compounds and Hemodynamics

Application of the knowledge of the electrophysiologic, negative inotropic, and vascular smooth muscle effects of calcium inhibitory compounds in isolated tissue preparations to normal and diseased animals or human patients is difficult. The net hemodynamic effects of calcium inhibitory compounds vary considerably depending upon their diverse pharmacologic activities independent of Ca^{2+} inhibition, their dose response characteristics, and their ability to initiate or inhibit central nervous system reflex mechanisms.

The administration of calcium inhibitory compounds to anesthetized or conscious animals and humans with a variety of cardiac diseases is associated with a dose dependent negative inotropic effect and decrease in myocardial oxygen consumption (88, 89, 95, 112, 114, 158, 159, 160). In general and at appropriate dosages, indexes of myocardial function including left ventricular pressure, left ventricular end diastolic pressure, left ventricular stroke work, the first derivative of left ventricular pressure (dP/dt_{max}), maximum velocity of contraction (V_{max}), and a diastolic relaxation constant are decreased, whereas coronary sinus oxygen saturation, left ventricular end systolic diameter, and cardiac output are increased (95, 114, 159, 161, 162, 163). Heart rate changes are unpredictable (161, 162). Various review articles have reported that calcium inhibitory compounds cause an increase, decrease or no change in heart rate (4, 6, 95, 131, 160). As the dose of calcium inhibitory compound increases, however, tachycardia generally develops (161). The hemodynamic changes described can be related to the ability of calcium inhibitory compounds to produce peripheral vasodilation (afterload reduction) and the interplay between direct drug mediated and reflex events. Calcium inhibitory compounds demonstrate little or no effect on venous capacitance. The administration of nifedipine into the left coronary artery of normal human volunteers, for example, produces decreases in dP/dt_{max}, maximum velocity of contraction, and left ventricular stroke work (161). In contrast, the intravenous administration of nifedipine into the intact circulation of conscious dogs or normal human volunteers fails to produce major changes in left ventricular stroke work, dP/dt_{max} or mean aortic blood pressure since vasodilation is reflexly counteracted by increases in cardiac output and heart rate (159, 162). Studies conducted in human patients with congestive heart failure indicate that the direct negative inotropic and chronotropic effects of nifedipine are masked by reflex sympathetic activation (163). The hemodynamic effects of verapamil are reportedly more variable than those elicited by nifedipine. Studies conducted in dogs and humans reveal inotropic and vasodilator effects similar to those of nifedipine except that reflex tachycardia is less likely to occur (95, 158, 159, 164). The reduced likelihood of verapamil induced tachycardia is attributed to a more prominent direct negative chronotropic effect (164). Heart rate effects similar to those reported for verapamil have been reported to occur after the intravenous administration of diltiazem to human patients and in dogs administered verapamil, diltiazem, and 2-substituted aminoindenes (52, 95, 140, 165).

The side effects reported in patients receiving calcium inhibitory compounds are an extension of the hemodynamic activity of this group of compounds and generally relate to drug potency and dilation of regional vascular beds (56, 131, 160). Depending upon dose, postural hypotension, dizziness, headache, AV conduction disturbances, pulmonary edema, heart failure and constipation can occur (6, 131, 166). Cardiac arrhythmias are a potential problem and if already present, could become more severe (161). The majority of these side effects are most frequently reported following nifedine administration although a greater long-term experience with all the calcium inhibitory compounds is required before a definite conclusion can be made regarding their safety. Despite the limited knowledge concerning the usage of calcium inhibitory compounds in patients, several authors have adopted general policies regarding their administration. It is suggested that calcium inhibitory compounds not be used or administered with caution to patients with significant SA or AV node disease, left ventricular outflow obstruction, low systolic blood pressure, paroxysmal nocturnal dyspnea or orthopnea (166). This list of contraindications is certain to increase as the calcium inhibitory compounds become more popular.

In conclusion, calcium membrane fluxes and variations in intracellular calcium activity play a pivotal role in maintaining normal smooth muscle and myocardial cell function. The slow inward current (I_{si}) carried principally by Ca^{2+} serves as the mechanism whereby extracellular Ca^{2+} can enter the myocardial cell to initiate or trigger excitation-contraction coupling. Although the mechanism(s) of calcium inhibitory compounds are disputed, the net effect of their administration is a reduction in intracellular calcium activity leading to a variety of beneficial and potentially deleterious effects. Because of the ability of calcium inhibitory compounds to regulate calcium activity and, therefore, a variety of cellular functions, they are potentially beneficial as therapy for angiospastic angina (167, 168), angina due to mechanical coronary artery occlusion (platelet aggregation) (169), cardiac arrhythmias (126), arterial hypertension (170), acute or chronic left ventricular failure (166), myocardial ischemia and infarction (158, 163), myocardial salvaging (cardioplegia) (153), cardiomyopathy (171) and varlous vasospastic syndromes (172). Compounds demonstrating calcium inhibitory activity require further development and refinement before current understanding of the excitation-contraction coupling process and the total usefulness of this group of compounds as therapeutic agents can be fully realized.

Literature Cited

1. Fleckenstein, A. "Calcium and the Heart", Academic Press, New York, 1971, pp 135-188.
2. Fleckenstein, A. Annu. Rev. Pharmacol. Toxicol. 1977, 17, 149-166.
3. Church, J.; Zsoter, T. T. Can. J. Physiol. Pharmacol. 1980, 58, 254-264.
4. Henry, P.D. "Trace Metals in Health and Disease", Raven Press, New York, 1979, pp 227-233.
5. Zsoter, T. T. Am. Heart J. 1980, 99, 805-810.
6. Henry, P. D. Am. J. Cardiol. 1980, 46, 1047-1058.
7. Nayler, W. G.; Poole-Wilson, P. Basic Res. Cardiol. 1981, 76, 1-15.
8. Fabiato, A.; Fabiato, F. Annu Rev. Physiol. 1974, 41, 473-484.
9. Langer, G. A. J. Mol. Cell. Cardiol. 1980, 12, 231-239.
10. Ringer, S. J. Physiol. (Lond) 1882, 4, 29-42.
11. McDonald, T. F.; Dieter, P.; Wolfgang, T. Circ. Res. 1981, 49, 576-583.
12. New, W.; Trautwein, W. Pfluegers Arch. 1972, 334, 24-38.
13. Trautwein, W.; McDonald, T. F.; Tripath, O. Pfluegers Arch. 1975, 354, 55-74.
14. Fabiato, A.; Fabiato, F. Circ. Res. 1977, 40, 119-129.
15. Fabiato, A.; Fabiato, F. Ann. N. Y. Acad. Sci. 1978, 307, 491-522.
16. Langer, G. A.; Frank, J. S. J. Cell. Biol. 1972, 54, 441-455.
17. Langer, G. A.; Frank, J. S. Am. J. Physiol. 1979, 237, H-239-H246.
18. Fabiato, A., Fabiato, F. Nature 1979, 281, 146-148.
19. Thorens, S.; Endo, M. Proc. Japan Acad. 1975, 51, 473-478.
20. Frank, J. S.; Langer, G. A.; Nudd, L. M.; Seraydarian, K. Circ. Res. 1977, 41, 702-714.
21. Langer, G. A.; Frank, J. S.; Philipson, K. D. Circ. Res. 1981, 49, 1289-1299.
22. Beeler, G. W. Jr.; Reuter, H. J. Physiol. 1970, 207, 211-229.
23. Gibbons, W. R.; Fozzard, H. A. J. Gen. Physiol. 1975, 65, 367-384.
24. Reuter, H. J. Physiol. 1967, 192, 479-192.
25. Wollenberger, A.; Will, H. Life Sci. 1978, 22, 1159-1178.
26. Wray, H. L.; Gray, R. R. Biochim. Biophys. Acta 1977, 461, 441-459.

27. Cheung, W. Y. Science 1980, 207, 19-27.
28. Katz, S.; Remtulla, M. A. Biochem. Biophys. Res. Commun. 1978, 83, 1373-1379.
29. Bilezikjian, L. M.; Kranias, E. G.; Potter, J. D.; Schwartz, A. Fed. Proc. 1980, 1663, Abstract.
30. Bilezikjian, L. M.; Kranias, E. G.; Potter, J. D.; Schwartz, A. Circ. Res. 1981, 49, 1356-1362.
31. Collins, J. H.; Kranias, E. G.; Reeves, A. S.; Bilezikjian, L.M.; Schwartz, A. Biochem. Biophys. Res. Commun. 1981, 99, 796-803.
32. LePeuch, C. J.; Haiech, J.; Demaille, J. G. Biochemistry 1979, 18, 5150-5157.
33. Glitsch, H. G.; Reuter, H.; Scholz, H. J. Physiol. (Lond) 1970, 209, 25-43.
34. Reuter, H. Circ. Res. 1974, 34, 599-605.
35. Coraboeuf, E.; Deroubaix, E.; Hoerter, J. Supp.l, Circ. Res. 1976, 38, I92-I98.
36. Diamond, J. M.; Wright, E. M. Annu. Rev. Physiol. 1969, 1, 581-646.
37. Kohlhardt, M.; Bauer, B.; Krause, H.; Fleckenstein, A. Pfluegers Arch. 1972, 335, 309-322.
38. Payet, M. D.; Schanne, O. F.; Ruiz-Ceretti, E. J. Mol. Cell. Cardiol. 1980, 12, 635-638.
39. Rosenberger, L.; Triggle, D. J. "Calcium in Drug Action"; Weiss, G.B., Ed.; Plenum Press: New York, 1978; 3-31.
40. Sanguinetti, M. C.; West, T. C. J. Pharmacol. Exp. Ther. 1981, 219, 715-722.
41. Bristow, M. R.; Green R. D. Eur. J. Pharmacol. 1981, 45, 267-279.
42. O'Hara Naoki; Ono H.; Oguro K,; Hashimoto, K. J. Cardiovasc. Pharmacol. 1981, 3, 251-268.
43. Kaufmann, R. Munch. Med. Wochenschr. 199 Suppl 1977, 1, 6-11.
44. Nagao, T, Sato, M, Nakajima, H, Kiyomoto, A. Chem. Pharm. Bull. 1973, 21, 92-97.
45. Fleckenstein, A.; Tritthart, H. S.; Doring, H. J.; Byon, Y. K. Arzneim. Forsch. 1972, 22, 22-33.
46. Kodama, I.; Hirata, Y.; Toyama, J.; Yamada, K. J. Cardiovas. Pharmacol. 1980, 2, 145-153.
47. Warltier, D. C.; Meils, C. M.; Gross, G. J.; Brooks, H. L. J. Pharmacola. Exp. Ther. 1981, 218, 296-302.
48. Ledda, F.; Mantelli, L.; Manzini, S.; Amerini, S.; Mugelli, A. J. Cardiovasc. Pharmacol. 1981, 3, 1162-1173.
49. Gmeiner, R.; Ng, C. K. J. Cardiovasc. Pharmacol. 1981, 3, 237-250.
50. Sutko, J. L.; Willerson, J. T. Circ. Res. 1980, 46, 332-343.
51. Siegal, M S.; Gliklich, J. I.; Mary-Robine, L.; Hoffmann, B. F. J. Pharmacol. Exp. Ther. 1979, 211, 606-614.

52. Piascik, M. F.; Piascik, M. T.; Witiak, D. T.; Rahwan, R. G. Can. J. Physiol. Pharmacol. 1979, 57, 1350-1358.
53. Rahwan, R. G.; Piascik, M. F.; Witiak, D. T. Can. J. Physiol. Pharmacol. 1979, 57, 443-460.
54. Zsoter, T. T. Am. Heart J. 1980, 99, 805-810.
55. Rahwan, R. G.; Faust, M. M.; Witiak, D. T. J. Pharmacol. Exp. Ther. 1977, 201, 126-137.
56. Sugiyama, S.; Kitazawa, M.; Kotaka, K.; Miyazaki, Y.; Ozawa, T. J. Cardiovasc. Pharmacol. 1981, 3, 801-806.
57. van Breemen, C.; Hwang, O.; Meisheri, K. D. J. Pharm. Exp. Ther. 1981, 218, 459-463.
58. Walus, K. M.; Fondacaro, J. D.; Jacobson, E. D. Circ. Res. 1981, 48, 692-700.
59. Wiggins, J. R. Circ. Res. 1981, 49, 718-725.
60. Dangman, K. H.; Hoffman, B. F. Am. J. Cardiol. 1980, 46, 1059-1067.
61. Danilo Jr, P.; Hordof, A. J.; Reder, R. F.; Rosen, M. R. J. Pharm. Exp. Ther. 1980, 213, 222-227.
62. Rosen, M. R.; Ilvento, J. P.; Gelband, H.; Merker, C. J. Pharm. Exp. Ther. 1974, 189, 414-422.
63. Ehara, T.; Kaufman, R. J. Pharmacol. Exp. Ther. 1978, 207, 49-55.
64. Reuter, H. Annu. Rev. Physiol. 1979, 41, 413-424.
65. Eckert R.; Tillotson D. L.; Brehm P. Fed. Proc. 1981, 40, 2226-2232.
66. Ito, S.; Surawicz B. Am. Physiol. Soc. 1981, H139-H144.
67. Lee, C. O.; Fozzard, H. A. Am. Physiol. Soc. 1979, C156-C165.
68. Colatsky, T J.; Hogan, P. M. Circ. Res. 1980, 46, 543-552.
69. Isenberg, G. Pfluegers Arch. 1977, 374, 77-85.
70. Siegal, M. S.; Hoffman, B. F. Circ. Res. 1980, 46, 227-236.
71. Noble, D.; Tsien, R. W. J. Physiol. 1968, 195; 185-214.
72. DiFrancesco, D. J. Physiol. 1981, 314, 377-393.
73. DiFrancesco, D.; Ohba, M.; Ojeda, C. J. Physiol. 1979, 297, 135-162.
74. Brown, H.; DiFrancesco, D. J. Physiol. 1980, 308, 331-351.
75. Singer, D. H.; Baumgarten, C. M.; Teneick, R. E. Prog. in Cardiovascl. Dis. 1981, 24, 97-156.
76. Bredikis, J.; Bukauskas, F.; Veteikis, R. Circ. Res. 1981, 49, 815-820.
77. DeMello, W. C. J. Physiol. 1976, 250, 171-197.
78. Watts, J. A.; Koch, C. D.; LaNoue, K. F. Am. Physiol. Soc. 1980, H909-H916.
79. Rosen, M R.; Danilo Jr, P. Circ. Res. 1980, 46, 117-124.
80. Vassalle, M.; Mugelli, A. Circ. Res. 1981, 48, 618-631.
81. Cranefield, P. F.; Aronson, R. S. Circ. Res. 1974, 34, 477-481.

82. Ten Eick, R. E.; Baumgarten, C. M.; Singer, D. H. Prog. in Cardiovascl. Dis. 1981, 24, 157-188.
83. Wit, A. L.; Cranefield, P. F.; Gadsby, D. C. Circ. Res. 1981, 49, 1029-1042.
84. Bayer, R.; Henekes, R.; Kaufman, R.; Mannhold, R. Naunyn-Schmiedeberg's Arch. Pharmacol. 1975, 290, 49-68.
85. Satoh, K.; Yanagisawa, T.; Taira, N. J. Cardiovasc. Pharmacol. 1980, 2, 309-318.
86. Endoh, M.; Yanagisawa, T.; Taira, N. Naunyn-Schmiedeberg's Arch. Pharmacol. 1978, 302, 235-238.
87. Gutovitz, A. L.; Cole, B.; Henry, P. D.; Sobel, B. E.; Roberts R. Circulation 1977, 56, Suppl III, 111-179.
88. Raschack, M. Naunyn-Schmiedeberg's Arch. Pharmacol. 1976, 294, 285-291.
89. Raschak, M. Arzneim. Forsch (Drug Res.), 1976, 26, 1330-1333.
90. Bayer, R.; Rodenkirchen, R.; Kaufman, R.; Lee, J. H.; Hennekes R. Naunyn-Schmiedeberg's Arch Pharmacol, 1977, 301, 29-37.
91. Cranefield, P. F.; Aronson, R. S.; Wit, A. L. Circ. Res. 1974, 34, 204-213.
92. Nabata, H. Japan J. Pharmacol. 1977, 27, 239-249.
93. Kass, R. S.; Tsien, R. W. J. Gen. Physiol. 1975, 66, 169-192.
94. Nawrath, H.; Ten Eick, R. E.; McDonald, T. F.; Trantwein, W. Circ. Res. 1977, 40, 408-414.
95. Gibson, R.; Driscoll, D.; Gillette, P.; Hartley, C. Dev. Pharmacol. Ther. 1981, 2, 104-116.
96. Kawai, C.; Konishi, T.; Matsuyama, E.; Entman, M. L. "Adalat. New Experimental and Clinical Results"; Excerpta Medica: Amsterdam, 1979, 5-6.
97. Narimatsu, A.; Taira, N. Naunyn-Schmiedeberg's Arch. Pharmacol. 1976, 294, 169-177.
98. Ono, H.; O'Hara, N. J. Cardiovas. Pharmacol. 1981, 3, 446-454.
99. Taira, N.; Narimatsu, A. Naunyn-Schmiedeberg's Arch. Pharmacol. 1975, 290, 107-112.
100. Zipes, D. P.; Fischer, J. C. Circ. Res. 1974, 34, 184-92.
101. Ono, H.; Himori, N.; Taira, N. Tohoku. J. Exp. Med. 1977, 121, 383-390.
102. Wit, A. L.; Cranfield, P. F. Circ. Res. 1977, 35, 413-425.
103. Katzung, B. G.; Morgenstern, J. A. Circ. Res. 1977, 40, 105-111.
104. Cranefield, P. F. Futura, 1975.
105. Wit, A. L.; Cranfield, P. F. Circ. Res. 1976, 38, 85-98.
106. Cranefield, P. F. Circ. Res. 1977, 41, 415-423.
107. Hordof, A. J.; Edie, R.; Malm, J. R. Hoffman, B. F.; Rosen, M. R. Circulation 1976, 54, 774-779.
108. Mary-Rabine, L.; Hardof, A. J.; Danilo, P. Jr; Malm, J.R.; Rosen, M.R. Circ. Res. 1980, 47, 267-277.

109. Elharrar, V.; Gaum, W. E.; Zipes, D. P. Am. J. Cardiol. 1977, 39, 544-549.
110. Dersham, G. H.; Han, J. J. Pharmacol. Exp. Ther. 1981, 216, 261-264.
111. Mangiard, L. M.; Hariman, R. J.; McAllister Jr, R. G.; Bhargava, V.; Surawicz, B.; Shabetas, R. Circulation 1978, 7, 366-372.
112. Newman, R. K.; Bishop, V. S.; Peterson, D. F.; Leroux, E. J.; Horwitz, L. D. J. Pharmacol. Exp. Ther. 1977, 201, 723-730.
113. Piascik, M. F.; Rahwan, R. G.; Witiak, D. T. J. Pharmacol. Exp. Ther. 1979, 210, 141-146.
114. Gross, R.; Kirchheim, H.; von Olshausen, K. Drug Res. 1979, 29, 1361-1368.
115. Ogawa, K.; Wakamasu, Y.; Ito T.; Suzuki, T.; Yamazaki, N. Drug Res. 1981, 31, 770-773.
116. Bush, L. R.; Li, Y. P.; Shlafer, M.; Jolly, S. R.; Lucchesi, B. R. J. Pharm. Exp. Ther. 1981, 218, 653-661.
117. Fujimoto, T.; Peter, T.; Hamamoto, H.; Mandel, W. J. Am. J. Card. 1981, 48, 851-857.
118. Hattori, S.; Weintraub, W. W.; Agarwal, J. B.; Bodenheimer, M. M.; Banka, V. S.; Helfant, R. H. Am. J. Cardiol. 1980, 45, 485, Abstract.
119. Henry, P. D. "Ischemic Myocardium and Antianginal Drugs"; Winbury M. M., Abiko Y., Eds.; Raven Press: New York, 1979; 143-153.
120. Millard, R W. Chest 1980, 78, 193-199.
121. Peng, C.; Kane, J. J.; Straub, K. D.; Murphy, M. L. J. Cardiovas. Pharmacol. 1980, 2, 45-54.
122. Selwyn, A. P.; Welman, E.; Fox, K.; Horlock, P.; Pratt, T.; Klein, M. Circ. Res. 1979, 44, 16-23.
123. Sherman, L. G.; Liang, C.; Boden, W. E; Hood, W. B. Circulation 1979, 29, 60, Suppl II, Abstract.
124. Weishaar, R.; Ashikawa, K.; Bing, R. J. Am. J. Cardiol. 1979, 43, 1137-1143.
125. Sperelakis, N.; Schneider, J. A. Am. J. Cardiol. 1976, 37, 1079-1085.
126. Kirkler, D. M.; Rowland, E., "Calcium Antagonismus"; Fleckenstein A., Roskamm H, Eds.; Springer-Verlag: Berlin, 1980; 55-61.
127. El-Sherif, N.; Lazzarra, R. Circulation 1979, 60, 605-615.
128. Fandacaro, J. D.; Han, J.; Yoon, M. S. Am. Heart J. 1978, 98, 81-86.
129. King, R. M.; Zipes, D. P.; DeBnicoll, A. Circulation 1974, 50, Suppl III, 111-183, Abstract.
130. Schamroth, L. Cardiovasc. Res. 1971, 5, 419-424.
131. Ellrodt, G.; Chew, C. Y. C.; Singh, B. N. Circulation 1980, 62, 669-679.
132. Pickering, T. G.; Goulding, L. Br. Heart Jr. 1978, 40, 851-855.

133. Singh, B. N.; Elldrodt, G.; Peter, C. Drugs 1978, 15, 169-197.
134. Waxman, H. L.; Myerberg, R. J.; Appel, R, Sung, R. J. Am. J. Cardiol. 1980, 45, 482, Abstract.
135. Spurrell, R. A. J.; Krikler, D. M.; Santon, E. Br. Heart J. 1974, 36, 256-264.
136. Lynch, J. J.; Rahwan, R. G.; Witiak, D. T. J. Cardiovasc. Pharmacol. 1981, 3, 49-60.
137. Gotsman, M. S.; Lewis, B. S.; Bakst, A.; Mitha, A. S. Afr. Med. J. 1972, 46, 2017-2019.
138. Hashimoto, K.; Taira, N.; Chiba, S.; Hashomoto Jr., K.; Endoh, M.; Kokobun, M.; Kokobun, H.; Iijima, T.; Kimura, T.; Kubota, K.; Oguro, K. Arzneim. Forsch. 1972, 22, 15-21.
139. Adelstein, R. S.; Hathaway, D. R. Am. J. Cardiol. 1979, 44, 783-787.
140. Franklin D,; Millard, R. W; Nagao, T. Chest 1980, 78, 200-204.
141. Mannhold, R.; Zierden, P.; Bayer, R.; Rodenkirchen, R.; Steiner, R. Drug Res. 1981, 31, 773-780.
142. Himori, N.; Ono, H.; Taira, N. Jpn. J. Pharmacol. 1976, 26, 427-435.
143. Guazzi, M.; Olivari, M. T.; Polese, A.;Fiorentini, C.; Magrini, F.; Moruzzi, P. Clin. Pharmacol. Ther. 1977, 22, 528-532.
144. Ito, Y.; Kuriyama, H.; Suzuki, H. Br. J. Pharmacol. 1978, 64, 503-510.
145. Schmier, J.; Bruckner, V. B.; Mittman, V.; Wirth, R. K. The Third International Adalat Symposium. Amsterdam: Excerpta Medica, 1976, 42-9.
146. Schmier, J.; VanAckern, K.; Bruckner, U. "The First Nifedipine Symposium"; Hashimoto, K.; Kimura, E.; Kobayashi, T., Eds.; Tokyo Press: Tokyo, 1975, 45-52.
147. Zsoter, T. T.; Wolchinsky, C.; Henein, N. F.; Hol, C. Cardiovas. Res. 1977, 11, 353-357.
148. Nagao, T.; Matlib, M. A.; Franklin, D.; Millard, R. W.; Schwartz, A. J. Mol. Cell. Cardiol. 1980, 12, 29-43.
149. Nayler, W. G.; Grau, A.; Slade, A. Cardiovasc. Res. 1976, 10, 650-662.
150. Smith, H. J.; Goldstein, R. A.; Griffith, J. M.; Kent, K. M.; Epstein, S. E. Circulation, 1976, 54, 629-635.
151. Cohen, L.; Gilula, Z.; Meier, P.; Lazaron, B.; Herbstman, D. J. Cardiovasc. Pharmacol. 1981, 3, 581-597.
152. Magee, P. G.; Flaherty, J. T.; Bixler, T. J.; Glower, D.; Gardner, T. J.; Burley, B. H.; Goh, V. L. Circulation. 1979, 60, Suppl I, 151-157.
153. Nayler, W. G.; Ferrari, R.; Williams, A. Am. J. Cardiol. 1980, 46, 242-248.
154. Shen, A. C.; Jennings, R. B. Am. J. Pathol. 1972, 67, 417-440.

155. Henry, P. D. "Ischemic Myocardium and Antianginal Drugs"; Winbury M. M.; Abiko Y., Eds.; Raven Press: New York, 1979, 143-153.
156. Vaghy, P. L.; Bor, P.; Szekeres, L. Biochem. Pharacol. 1980, 29, 1385-1389.
157. Bor, P.; Pataricza, J.; Vaghy, P. L.; Szekeres, L. Proc. Int. Union Physiol. Sci. 1980, 14, 333.
158. Atterhog, J. E.; Ekelung, L. G. Eur. J. Clin. Pharmacol. 1975, 8, 317-322.
159. Carlens, P. J. Cardiovasc. Pharmacol. 1981, 3, 1-10.
160. Stone, P. H.; Antman, E. M.; Muller, J. E.; Braunwald, E. Ann. Intern. Med. 1980, 93, 886-904.
161. Hollman, W.; Rost, R.; Liesen, H.; Emirkanian, O. "The Second Int. Adalat Symposium"; Lochner W.; Braasch W.; Kroneberg G., Eds.; Springer-Verlag: New York, 1976, 243-247.
162. Lichtlen, P. "The First Nifedipine Symposium"; Hashimoto, K; Kimura, E; Kobayashi, T, Eds.; Tokyo Press: Tokyo, 1975, 114-119.
163. Matsumoto, S.; Ito, T.; Sada, T.; Takanashi, M.; Su, K. M.; Ueda, A.; Okabe, F.; Sato, M.; Sekine, I.; Ito, Y. Am. J. Cardiol. 1980, 46, 476-480.
164. Lewis, B. S.; Mitha, A. S.; Gofsman, M. S. Cardiology 1975, 60, 366-376.
165. Kusukawa, R.; Kinoshita, M.; Shimono, Y.; Tomonaga, G.; Hosino, T. Arzneim. Forsch. 1977, 27, 878-887.
166. Epstein, S. E.; Rosing, D. R. Circulation 1981, 64, 437-441.
167. Freedman, B.; Dunn, R. F.; Richmond, D. R.; Kelly, D. T. Circulation 1979, 60, Suppl II, 249, Abstract.
168. Hosada, S.; Kimura, E. "The Third International Adalat Symposium"; Jatene A. D.; Lichtlen P. R., Eds.; Excerpta Medica: Amsterdam, 1976, 174, 195-199.
169. Kaltenbach, M. "The First Nifedipine Symposium"; Hashimoto K.; Kimura E.; Kobayashi T., Eds.; Tokyo Press: Tokyo, 1975, 189, 126-135.
170. Blaustein M. P. Am. J. Physiol. 1977, 232, C165-C173.
171. Kaltenbach, M.; Hopf, R.; Kober, G.; Bussmann, W. D.; Keiler, M.; Petersen, Y. Br. Heart J. 1979, 42, 35-42.
172. Webb, R. C.; Bohr, D. V. Prog. Cardiovasc. Dis. 1981, 24, 213-242.

RECEIVED June 16, 1982.

4

Control of Intracellular Calcium in Smooth Muscle

E. E. DANIEL, A. K. GROVER, and C. Y. KWAN

McMaster University Health Science Centre, Department of Neurosciences, Hamilton, Ontario, Canada L8N 3Z5

Control of intracellular calcium in smooth muscle cells is essential for control of the contractile state of the cells. Elevation of intra-cellular Ca^{2+} is achieved by opening membrane channels for transport of Ca^{2+} down its electrochemical gradient. The nature of these channels and their interaction with drugs including Ca^{2+} antagonists are briefly considered. In some smooth muscles, elevation of intra-cellular Ca^{2+} can occur by release of Ca^{2+} sequestered in the cell; the role of this process in initiation of contraction is unclear. Lowering of intracellular Ca^{2+} to maintain or allow relaxation may utilize a variety of mechanisms. The major focus of this article is to summarize and evaluate the data showing that the plasma membrane plays a major role in lowering intracellular Ca^{2+}. This evidence has been obtained by isolating and purifying plasma membrane vesicles and studying their transport properties. They possess a vectorial ATP-dependent Ca^{2+} transport system capable of accumulating 1000-fold or higher Ca^{2+} gradients. The properties of this system are described. Plasma membrane vesicles also possess a Na^{+} - Ca^{2+} exchange system and ATP-independent binding mechanisms. The direction of future research to evaluate the contributions of these systems to control of intra-cellular Ca^{2+} is discussed.

Control of intracellular Ca^{2+}, $[Ca^{2+}_i]$, in uterine and other smooth muscles is essential for control of tension production. Studies of smooth muscles with plasma membranes damaged by glycerol (1) or non-ionic detergents (2,3,4) or of isolated contractile protein from smooth muscle (5,6,7) all suggest that with less than 10^{-7} M Ca^{2+}_i no active tension is produced while at 10^{-5} M Ca^{2+}_i or perhaps less, maximum active tension is produced.

0097-6156/82/0201-0073$06.00/0

Elevation of free Mg^{2+} increases the Ca^{2+}_i concentration requirement slightly (4). Intermediate concentrations lead to graded changes in tension. The obvious problem is how the smooth muscle cells can regulate Ca^{2+}_i over this range.

Elevation of $[Ca^{2+}_i]$

Increases in $[Ca^{2+}_i]$ are, at first glance, easy to achieve since the external Ca^{2+} concentration $[Ca^{2+}_i]$ is about 10^{-3} M, 100 to 10,000-fold higher than internal Ca^{2+} (< 10^{-7} M in relaxed muscle) and the transmembrane electrical gradient is -40 to -60mv, inside negative. Opening of an inward Ca^{2+} leak down the electrochemical gradient is thus an obvious way to increase Ca^{2+}_i. However, this leak must be capable of being opened and closed on command; the opening commands identified experimentally (8,9) are decrease in the transmembrane voltage gradient (voltage-dependent Ca^{2+} channels) or activation by occupation (by a neurotransmitter or hormone) of a receptor linked to a Ca^{2+} channel in the membrane (receptor operated Ca^{2+} channel). The latter can open without membrane depolarization (10,11), even accompanied by hyperpolarization (12). Whether they are also sometimes voltage-sensitive requires further study (see Bolton for review of these channels (9). Contraction believed to occur by Ca influx without depolarization by opening of receptor-operated channels, has been termed pharmacomechanical coupling; that believed to occur with depolarization by opening of voltage-dependent channels has been termed electro-mechanical coupling (10). In many (but not all) smooth muscles (see 8,9), voltage-operated Ca^{2+} channels undergo regenerative increases in conductance on partial depolarization like those to Na^+ in nerve and skeletal muscle; this results in Ca^{2+} spikes or action potentials (13,14,15,16). Voltage operated Ca^{2+} channels show inactivation; i.e., they do not remain open indefinitely (13-16). Furthermore, triggered by increase in Ca^{2+}_i near the membrane, there is often an increase in K^+ conductance, leading to hyperpolarization of the cell membrane and closure of Ca^{2+} channels (17-19). Since $[K^+_i]$ is usually > 150 mM and $[K^+_e]$ is about 5 mM, the electrochemical equilibrium for a membrane passing mostly K^+ currents occurs at about -90 mv. Hyperpolarization activated by elevation of internal Ca^{2+} is a widespread phenomenon in nature, occurring in many cell types (21,22). Either inactivation of Ca^{2+} channels on continued depolarization or their closure on repolarization can help terminate contraction or lead to refractoriness.

Not only are voltage-dependent transmembrane Ca^{2+} channels closed by inactivation and by Ca^{2+}_i induced hyperpolarization secondary to K^+-conductance increase, they are also affected by inhibitory neurotransmitters and hormones (see 9). Occupation of the membrane receptors for these agents opens conductance channels (for K^+ or Cl^- or both) or activates vectorial electrogenic, energy-dependent ion flow (active transport) which hyperpolarize the cell membrane and turn off voltage dependent Ca^{2+} channels. One would expect membrane hyperpolarization by any mechanism to be in- or less effective in reducing Ca^{2+} influx via receptor-operated than voltage-dependent channels; but this does not appear to have been systematically studied. Whether receptor-operated channels are inactivated with time or closed by any other mechanism than removal of the occupying agonist, is unclear.

Methods for Study of Ca Channels; Ca^{2+}-antagonists

The outlines of mechanisms for increasing Ca^{2+} influx by opening transmembrane Ca^{2+} channels have been worked out using electrophysiological measurements of transmembrane potentials (see Bolton, 9), attempting to measure transmembrane currents by voltage clamp (13-16) or manipulating external Ca^{2+} ion concentrations and transmembrane potentials (see Bolton, 9) while studying the effects of various neurotransmitters, hormones or other agents affecting the contractile state. Lately, these studies have been stimulated by the availability of "Ca^{2+}-antagonists", believed to act to block Ca channels, possibly by interacting with them (22,23). However, direct proof of their mode of action is missing in smooth muscle and their mechanisms are inferred from their selective effects on Ca^{2+} depolarization or Ca^{2+} spikes (13-16), the competitive nature of their interactions with Ca^{2+}_e in relation to these effects or to contractions (22,23) and from studies in which La^{3+}-resistant Ca^{2+} influx was measured (24,25). So far no one has isolated any Ca channel, identified it chemically, determined what substance can interact with it or reconstituted it in a lipid bilayer. Much evidence suggests a variety of modes of action of agents labelled Ca^{2+}-antagonists (23); eg. besides antagonism to Ca^{2+} entry, enhanced Ca^{2+} extrusion or sequestration or interference with Ca^{2+} actions. Thus the understanding of Ca channels in smooth muscle is based so far on physiological and pharmacological experiments and inferences from them; not on study of the physical- and biochemistry of Ca^{2+} channels. Their regulation by the Ca^{2+} regulatory protein, calmodulin, has not yet been analyzed.

Ca^{2+} Increase by Release of Sequestered Ca^{2+}

Surprisingly, perhaps, there appear to be alternate modes of increasing $[Ca^{2+}_i]$ in some smooth muscles in addition to opening Ca^{2+} channels; i.e., by release of Ca^{2+} sequestered in a fashion or at a site not allowing rapid and free exchange with the extracellular medium (see 26). It is usually inferred (27,28,29,30) that sites of sequestration are intracellular organelles because of poor availability of this Ca^{2+} to washout, to calcium chelating agents (eg. EGTA) or displacing agents (La^{3+}) in the external medium, but the possibility that this Ca^{2+} is tightly bound to the internal or even the external cell surface, possibly in sites of impeded exchange (eg. caveolae) has not been excluded (26). Mechanisms for increasing $[Ca^{2+}_i]$ by release of sequestered Ca^{2+} can be easily confused with pharmacomechanical coupling since depolarization is not a prerequisite for either; differentiation of release of sequestered Ca^{2+} from opening of receptor-operated channels for Ca^{2+} depends upon quantitative considerations of the ease of inhibition by removal of external Ca^{2+} or by addition of presumed Ca^{2+} channel antagonists (see 26).

Release of sequestered Ca^{2+} to support contraction often occurs only with higher doses of neurotransmitters, hormones or their analogs (28,29), is limited in that only one or a few contractile responses can be supported without renewal of the Ca^{2+} store (27-30), usually requires resupply of the Ca^{2+} store from the external medium rather than recycling from the internal medium (27-30). It may have mechanistic rather than physiological significance, representing an artifact of experimental arrangements which reveal a site of Ca^{2+} binding capable of release by large concentrations of agonists. The major question is whether release of sequestered Ca^{2+} plays a central role in excitation-contraction coupling.

Ca Extrusion and Sequestration

Inhibition of Ca^{2+} influx or release would not restore $[Ca^{2+}_i]$ to 10^{-7} M since any residual Ca^{2+} leak would be inward. Thus an extrusion or sequestering mechanism must be employed to lower Ca^{2+}_i. A sequestering mechanism which cyclically replenishes an internal store of sequestered Ca^{2+} after its release for contraction would appear initially to be an efficient way to provide for relaxation and contraction of smooth muscle. However, it implies a non-existent isolation of the internal from the external fluid, an unlikely directional flow of Ca^{2+} to and from sequestering sites or a geographic proximity of these sites to contractile proteins. In fact, there is clear evidence that an active (ATP-dependent) Ca^{2+} extrusion mechanism exists in the plasma membrane and that it is probably capable of removing Ca^{2+}_i and creating the necessary inward gradient of 1,000 to 10,000-fold (26,31,32,33,34). This has been established by isolation of highly purified plasma membrane vesicles and studying their transport properties (31,32). The nature of the evidence is the main focus of this paper.

Three major areas of controversy exist. First, is there an additional Ca^{2+} transport or sequestering mechanism into endoplasmic reticulum or mitochondria and if so, what is its importance relative to that of the plasma membrane pump? Second, is there a second plasma membrane extrusion system for Ca^{2+} which couples Ca^{2+} extrusion to Na^+ influx using the energy of the Na^+ inward gradient created by the well known Na^+ - K^+ pump. Third, is it the endoplasmic reticulum or the plasma membrane from which sequestered Ca^{2+} is released by some agonists (see above)? This paper will address only the first two since we are currently just beginning study of the third question.

An operational model for control of Ca^{2+}_i in smooth muscle would involve the components identified in Figure 1. An attempt has been made to suggest what processes and inter-relationships are established, which are probable, which are uncertain or even unlikely.

Methods for Study of Ca^{2+} Transport and Binding by Isolated Membranes.

A standardized approach has been developed for arriving at a method for isolation of purified plasma membrane from each of various smooth muscles (31-41; see 26,33,34). It starts with a very careful dissection to remove all possible cells other than smooth muscle. Studies have shown that what appears to be a minor contamination by non-smooth muscle components (eg. fat cells or adventitia of blood vessels) may markedly affect the overall properties of the subsequent membrane fractions isolated (36). Since many smooth muscle organs contain two or more layers of muscle which often differ in their physiological (42), pharmacological (43) and isolated membrane properties (44), it is always desirable to separate muscle layers and to isolate their membranes individually. There are inevitably a few fibroblasts and nerves left in the most carefully dissected smooth muscle tissue; so far, no one has compared these properties of membranes from smooth muscle tissue to those from freshly isolated smooth muscle cells to determine if these residual contaminent membranes affect the properties of the membranes from tissues; we think not.

The isolated smooth muscle is usually homogenized using a Polytron at varying speed and duration to attain maximal yield of plasma membrane with minimal contamination by broken mitochondrial membranes. The mitochondria are separated by multiple applications of differential centrifugation. The resultant microsomal membranes are brought down by a high speed spin and then applied to a sucrose density gradient. In the first instances a continuous gradient is used and fractions are isolated and then protein contents and marker enzyme activities or binding noted (36,45). They include enzyme markers (35), ligands for membrane receptors (34,35), ligands for specific plasma membrane chemicals and digitonin affinity of plasma membrane cholesterol (46). Some of these we use are summarized in Table 1. On the basis of these observations a discontinuous sucrose gradient is constructed with the objective of minimum contamination of plasma membrane by other membranes with adequate yield. Figures 2 and 3 contain a typical example of these procedures and the resultant separation of membranes based on marker studies. Techniques have been developed to estimate the purity of plasma membrane and other fractions (35,36). The purity of plasma membrane in several of these vesicular fractions is estimated in Table 2 and physical

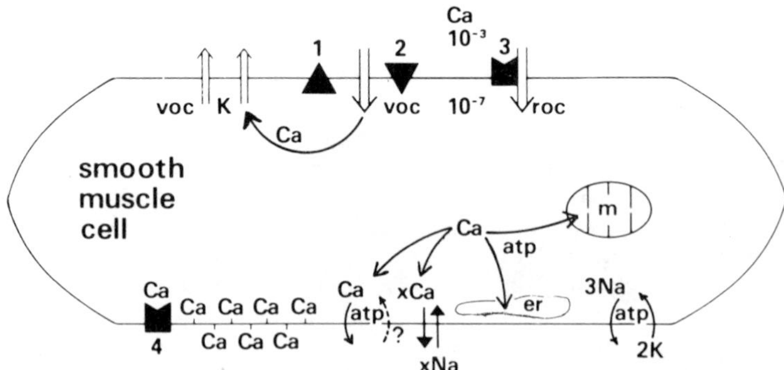

Figure 1. Schematic diagram of control of intracellular Ca^{2+} in smooth muscle cells. Key: Ca_0^{2+} (outside), 10^{-3} M *and Ca_i^{2+} (inside), 10^{-7}* M.

At top are shown (from left to right): Voltage-operated K^+ channel (VOC-K) which tends to hyperpolarize membrane Ca_i^{2+}-operated K channel (also hyperpolarizing); 1. Receptor for inhibitory (hyperpolarizing) neurotransmitter; VOC-voltage-operated Ca^{2+} channel; 2. Receptor for excitatory (depolarizing) neurotransmitter; 3. Receptor for neurotransmitter, etc. which opens receptor-operated Ca^{2+} channel. Ca inside the cell may be sequestered in mitochondria (M) or endoplasmic reticulum (ER) by ATP-dependent processes. Of more importance (bottom) are ATP-dependent Ca^{2+} outward transport (with unknown stoichiometry and counter or co-ion transport) and possible Ca^{2+}–Na^+ exchange (with unknown stoichiometry and electrogenicity). This Ca^{2+}–Na^+ exchange and normal membrane potential and responses to conductance increases depend upon the maintenance of Na^+, K^+ and voltage gradients by the ATP-dependent, electrogenic Na pump. Ca^{2+} can be bound to membranes inside or out and may be released by neurotransmitter combination with the receptors (4) which may be the same as the receptor at 3.

TABLE I

Markers for Smooth Muscle Membranes

Plasma Membrane:	References to Use:
5'-nucleotidase	35, 37
Phosphodiesterase I	40
K^+-Na^+ activated Mg^{2+} ATPase (ouabain-sensitive)	45
K^+-activated phosphatase (ouabain-sensitive)	37
Specific Binding of Agonists or Antagonists: eg. Oxytocin, PGs, adrenergic, muscarinic	35
Digitonin binding	46
Wheat germ agglutinin binding	35
Endoplasmic Reticulum:	
NADPH cytochrome c reductase	35, 37, 40
Mitochondria: Inner Membrane	
Cytochrome c oxidase	36
Succinate cytochrome c reductase	35
Mitochondria: Outer Membrane	
Rotenone-insensitive NADH cytochrome c reductase	35
Monoamine oxidase	35
Membrane:	
Glucoronidase	35

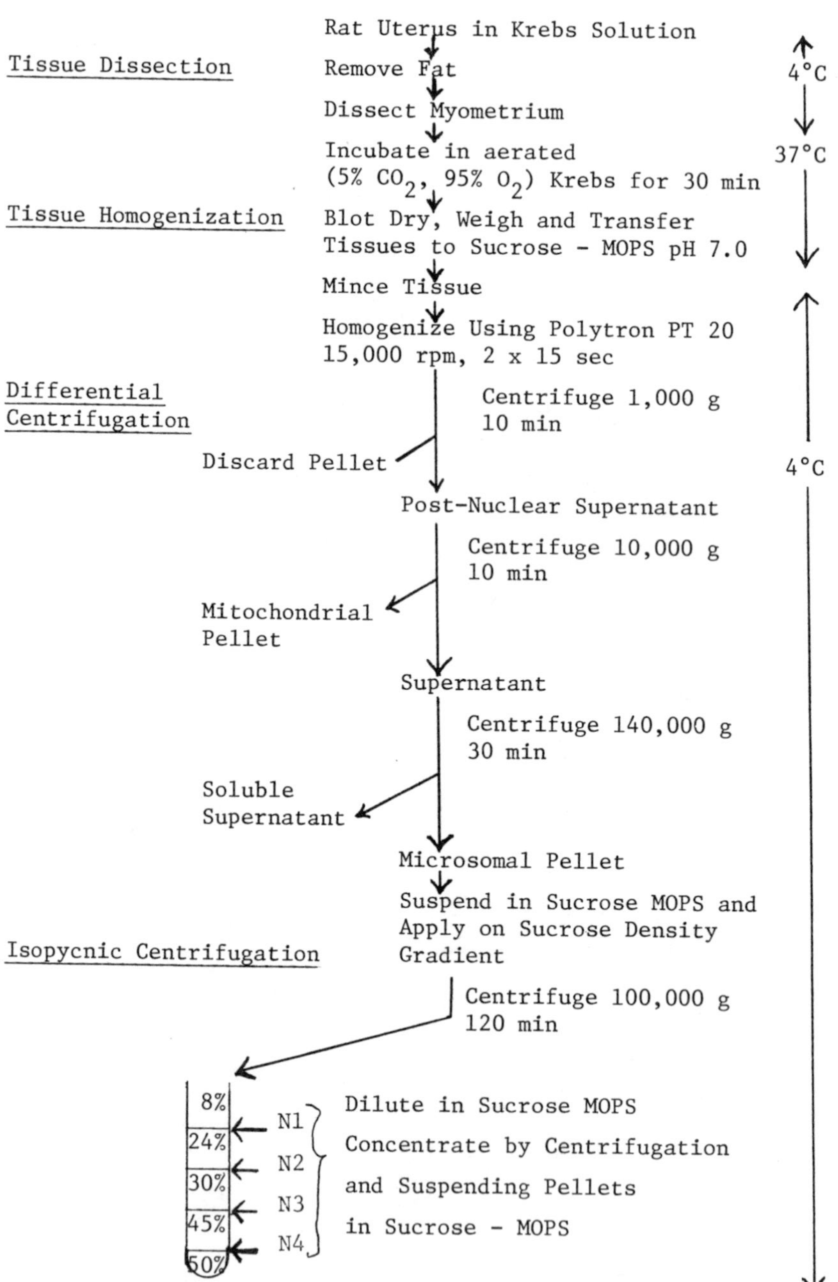

Figure 2. Scheme for isolation of purified plasma membrane fractions from longitudinal muscle of myometrium from estrogenized rats. Variations of the basic approach are used in various smooth muscles.

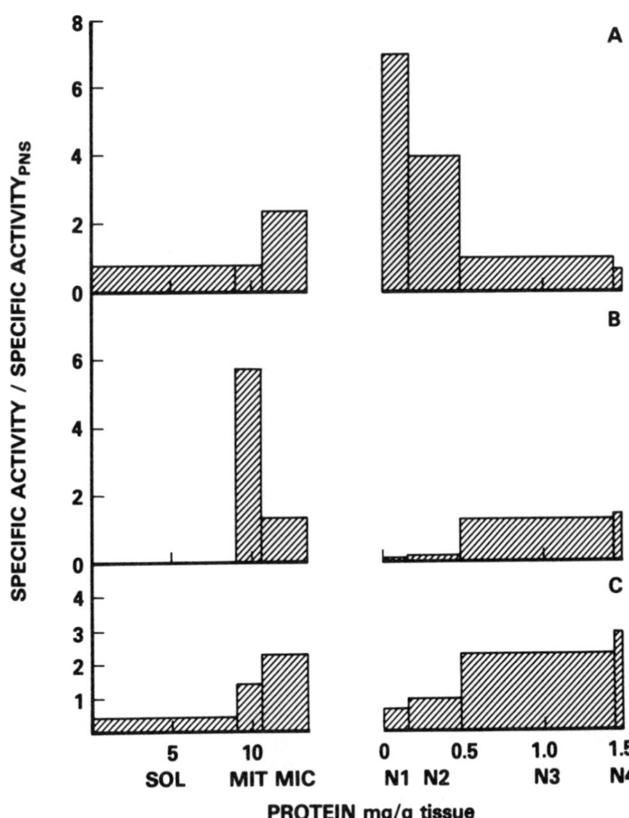

Figure 3. DeDuve plots of relative specific activities (specific activity/specific activity in PNS) in various fractions. Key to fractions: PNS, postnuclear supernatant; SOL, soluble; MIT, mitochondria; MIC, microsomes; and N1, N2, N3, and N4, the discontinuous gradient fractions (see Figure 2). Key to activities: A, 5′-nucleotidase; B, succinate–cytochrome c reductase; and C, NADPH–cytochrome c reductase.

properties of our most purified plasma membrane fraction in Table 3.

Properties of Isolated Plasma Membranes

The Ca binding and transport properties of these membranes have been studied; both passive (ATP-independent) and active (ATP-dependent) binding occur; in addition, at least in vesicles in which the plasma membranes are oriented inside-out (estimated to be 40% in one case, 47), there is an ATP-dependent Ca^{2+} transport. The Ca^{2+} is accumulated inside these vesicles, and > 1,000-fold gradients of Ca^{2+} across the vesicles membrane have been estimated. Since a transvesicle membrane potential was not knowingly produced, the Ca^{2+} accumulations may represent the gradients possible with voltage operated channels open. On the other hand, these channels may have been damaged or inactivated during membrane isolation. Ca^{2+} accumulated inside properly oriented vesicles is slowly released when they are diluted into Ca-free or ^{45}Ca-free media; it is rapidly, almost instantaneously released when the Ca^{2+}-selective ionophores (A23187, ionomycin, or X537A) are added (31,35,36,39).

In unpublished studies we have found the Ca ATP-dependent transport by plasma membrane vesicles to be enhanced by calmodulin and inhibited by phenothiazines and to be stimulated by c-AMP dependent protein kinases. The mechanisms of action of these modulating systems and the physiological functions remain to be determined.

Na^{+} - Ca^{2+} Exchange

These plasma membrane vesicles, at least in the two cases examined to date (33,48), can exchange Ca^{2+} for Na^{+}; There is increased Ca^{2+} uptake when an outwardly directed Na^{+} gradient is formed across the vesicles and increased Ca^{2+} efflux when an inwardly directed Na^{+} gradient is formed. These effects could not be duplicated by K^{+}, Rb^{+} or Cs^{+} gradients, and Li^{+} gradients had only a slight ability to effect these changes. Furthermore, the Na^{+}-gradient supported Ca-uptake can also occur against a net outward Ca-gradient (unpublished data). The Na^{+} selective ionophore, monensin, or the Ca^{2+} selective ionophore, A23187, each abolished these effects. Thus, it is impossible to explain these results in terms of a competition between Na^{+} and Ca^{2+} for binding to anionic sites on the plasma membrane; rather the Na^{+} gradients must be providing the energy for driving Ca^{2+} in the opposite direction. This, of course, does not exclude an additional role played by binding interactions between Na^{+} or other cations and Ca^{2+}.

What role this exchange system plays in the intact cell, remains to be decided. Studies of intact cells *in vitro* suggest, not surprisingly, that for Ca^{2+}_{i}, Na^{+} - Ca^{2+} exchange cannot

Table II

Estimated Purity of Plasma Membrane[+] Fractions

		References
Rat Myometrium:	N_1 > 95%	31, 35, 46, 47
	N_2 ∿ 60%[++]	
Rat Mesenteric Artery:	F_1 70 - 80%	36
Rat Fundus:	F_2 70 - 80%	33, 40
Rat Mesenteric Vein:	F_1 70 - 80%	38
Rat Vas Deferens:	F_2 70 - 80%	33, 40
Canine Trachealis:	F_2 70 - 80%	37
Canine Gastric Corpus: (circular muscle)	F_3 ∿70 - 80%	39

[+]No fractions of endoplasmic reticulum were so far obtained.
> 50% purity.

[++]Most of the impurity in this fraction appears to be attached contractile protein (31, 46).

Table III

Physical Properties of Rat Myometrium Plasma Membrane

		References
Density:	1.08 - 1.03 g/ml	31
Size:	Smooth surface vesicles with cross-sectional diameters of 0.05 - 0.5 μm	
Trapping Volume:	1 - 3 μl/mg protein using sucrose and 4 - 5 μl/mg using inulin (unpublished)	47
Protein: Phospholipid Ratio:	∿ 1 (w/w)	unpublished
Orientation:	20% broken	47
	40% sealed rightside-out	
	40% sealed inside-out	

provide a complete control (49,50) but may participate in regulating it by affecting Ca fluxes (30,51).

In the isolated plasma membrane vesicles which we studied, the Na^+-pump was not activated (no ATP present) and the Na^+ gradients were shortlived (1 to 3 minutes). Also, these isolated vesicles may have no transmembrane potential difference unlike the intact cell; so the total electrochemical gradient was less. Furthermore, they probably have an increased leak of both Na^+ and Ca^{2+} through their membrane. The absence of ATP may have reduced the affinity of exchange sites for Ca^{2+}. All these uncertainties limit the conclusions about the potential role of Na^+ - Ca^{2+} exchange that should be drawn at this stage. Nevertheless, the maximum initial rate of Ca^{2+} transport (from a medium containing 40 μM free Ca^{2+}) driven by the Na^+ gradient was comparable to that effected by the ATP-dependent Ca^{2+} pump (from a medium containing 1 μM free Ca^{2+}) (34). This rate was obtained with a Na^+-gradient of about 20-fold. In the intact cell, the Na^+-gradient is probably about 10-fold and the internal Ca^{2+} concentration during relaxation is 0.1 μM; so this Na^+ - Ca^{2+} exchange system may operate rapidly only where internal Ca^{2+} concentrations are elevated.

A final judgment cannot be made until Na^+ - Ca^{2+} exchange has been studied under optimal conditions in the presence of a maintained Na^+ gradient in isolated plasma membranes and intact cells. The stoichiometry, electrogenicity, Ca^{2+}-gradient dependence, Na^+-gradient dependence, as well as influence of ATP and various regulating factors need to be determined. Variation between Na^+ - Ca^{2+} exchange in plasma membranes of different smooth muscles is to be expected, adding to present uncertainty.

A similar suggestion has been made that internal sequestration of Ca^{2+} by mitochondria occurs only at high internal Ca^{2+} concentrations, based on their low Ca^{2+} content in studies using x-ray diffraction analysis of sections rapidly frozen to minimize translocations (52,53) and on studies of the Ca^{2+} dependence for Ca^{2+} transport (54,55). In human myometrium, however, isolated mitochondria could transport Ca^{2+} from concentrations less than 1 μM (56).

Endoplasmic Reticulum and Regulation of Ca^{2+}_i

Based largely on analogy to striated muscle, it has become a dogma that the major site of Ca^{2+} sequestration and transport in smooth muscle is in the endoplasmic reticulum (ER). The best positive evidence for this consists of studies by electron probe of rapidly frozen muscles (52,53,57); so far only large blood vessels of guinea pig have been studied and in them Ca^{2+} was reported to be accumulated in ER under conditions when intracellular Ca^{2+} was increased by K^+-depolarization and even in relaxed muscle. This method has important limitations; it would not detect a Ca^{2+} transport function by plasma membrane since the Ca^{2+} is transported into the extracellular space. Also its ability to resolve Ca^{2+} bound to one surface of a biological membrane is not established. Finally no quantitative study of Ca^{2+} translocations during preparations of specimens is available. Other problems exist for application of this evidence to provide general support for the dogma. First, many smooth muscle cells, especially small arteries, veins, and some gut muscles contain a small volume of ER, insufficient to provide for accumulation of the necessary Ca^{2+}. Second, to function as a major functional site for Ca^{2+} accumulation during repetitive contraction cycles, ER must either release Ca^{2+} again after accumulation or in some way translocate it to the exterior. No direct evidence for either of these processes exists in fully studied smooth muscles (see above). Third, no one has been able to isolate a highly purified ER fraction with the appropriate Ca^{2+} transport properties. Instead, crude microsomal fractions containing a large amount (probably its major Ca-transporing component) of PM have been used. The observed Ca^{2+} transport properties were in most cases related to this membrane. However, Carsten and Miller (58) have isolated from bovine myometrium a membrane fraction apparently denser than PM and with the ability to accumulate Ca^{2+} in an ATP-dependent, oxalate-stimulated manner. We have found in one tissue, rat vas deferens (41), that in addition to plasma membrane there was a denser fraction enriched in NADPH-cytochrome c reductase which had similar Ca^{2+} transport properties. Thus, it seems probable that endoplasmic reticulum, at least in some smooth muscles, can accumulate Ca^{2+} and in muscles where present in sufficient amounts, may be a site of Ca^{2+} sequestration. Properties of these ER membrane fractions have not been studied in detail; also they contain much plasma membrane, making interpretations tentative. Development of highly purified ER fractions and study of their Ca^{2+} handling properties is an obvious goal for the next several years.

The Future Study of Isolated Membranes

The ultimate goal of studies of Ca^{2+} handling by isolated membrane system is to study under biochemically determined conditions the molecular events of excitation - contraction coupling. Such studies could settle all controversies if they are able to show that appropriate agonists open or close Ca^{2+} channels or release bound or sequestered Ca^{2+} under defined conditions with pure membrane fractions and that these events are sufficient to account for the events observed in the living cell. We have arrived at the stage of technical achievement with regard to membrane isolation and study that we can consider that these goals are within reach.

Acknowledgment

Various aspects of this work were supported by the Medical Research Council of Canada and the Ontario Heart Foundation. Expert technical assistance from Ms. J. Crankshaw is gratefully noted.

Literature Cited

1. Filo, R.S.; Bohr, D.F.; Ruegg, J.C. Science 1965, 147, 1581-1583.
2. Endo, M.; Kitozawa, T.; Yoki, S.; Iino, M.; Kakuta, Y. "Excitation-Contraction Coupling in Smooth Muscle", Eds. R. Casteels, T. Godfraind, J.C. Ruegg. Elsevier-North Holland Biomedical Press: Amsterdam, 1977; pp 199-209.
3. Cassidy, P.; Hoar, P.E.; Kerrick, W.G.L. Pflüger's Arch. Physiol. 1980, 387, 115-120.
4. Cornelius, F. J. Gen. Physiol. 1980, 25, 709-725.
5. Ebashi, S.; Endo, M. Prog. Biophys. Mol. Biol. 1968, 2, 351-385.
6. Marstain, S.B.; Trevett, R.M.; Walters, M. Biochem. J. 1980, 185, 355-365.
7. Hirata, M.; Nikawa, T.; Nonomura, Y.; Ebashi, J. J. Biochem. 1980, 87, 369-378.
8. Casteels, R. Chest 1980, 78, Suppl. 1, 151-156.
9. Bolton, J.B. Physiol. Rev. 1979, 59, 606-718.
10. Somlyo, A.P.; Somlyo, A.V. J. Pharmacol. Exp. Ther. 1968, 159, 129-145.
11. Droogmans, G.; Raeymaekers, L.; Casteels, R. J. Gen. Physiol. 1977, 70, 129-148.
12. Kitamura, K.; Kuriyama, H. J. Physiol. (Lond.) 1979, 293, 119-133.
13. Mironneau, J. J. Physiol. (Lond.) 1973, 233, 127-141.
14. Kao, C.Y.; McCollough, J.R. J. Physiol. (Lond.) 1975, 246, 1-36.
15. Inomata, H.; Kao, C.Y. J. Physiol. (Lond.) 1976, 255, 347-378.
16. Walsh, J.V.,Jr.; Singer, J.J. Am. J. Physiol. 1980, 239, C162-C174.
17. Vassort, G. J. Physiol. (Lond.) 1975, 252, 713-734.
18. Inomata, H., Kao, C.Y. J. Physiol. (Lond.) 1979, 297, 443-462.
19. Mironneau, J.; Saveneau, J.P. J. Physiol. (Lond.) 1980, 302, 411-425.
20. Walsh, J.V.,Jr.; Singer, J.J. Am. J. Physiol. 1980, 239, C182-C189.
21. Putney, J.W. Pharmacol. Rev. 1978, 30, 209-245.
22. Fleckenstein, A.; Nakayama, K.; Fleckenstein-Grün, G.; Byon, Y.K. "Coronary angiography and angina pectoris" Ed. P.R. Lichten; Publ. Sci. Group: Acton, MA, 1976; pp. 297-315.
23. Triggle, D.J.; Swamy, V.C. Chest 1980, 78, Suppl.
24. Triggle, C.R.; Swamy, V.C.; Triggle, D.J. Can. J. Physiol. Pharmacol. 1979, 57, 804-818.
25. Rosenberger, L.; Ticku, M.K.; Triggle, D.J. Can. J. Physiol. Pharmacol. 1979, 57, 333-347.
26. Daniel, E.E.; Crankshaw, D.; Kwan, C.Y. "Trends in Autonomic Pharmacology" Ed. S. Kalsner, Urban and Schwarzenberg Inc., Germany, 1979; pp. 443-484.

27. Deth, R.; Van Breemen, C. J. Membr. Biol. 1977, 30, 363-380.
28. Casteels, R.; Raeymaekers, L. J. Physiol. (Lond.) 1979, 294, 51-68.
29. Brading, A.F.; Sneddon, P. Brit. J. Pharmacol. 1980, 70, 229-240.
30. Van Breemen, C.; Aaronson, P.; Loutzenhiser, R.; Meisheri, K. Chest 1980, 78, Suppl. 157-165.
31. Grover, A.K.; Kwan, C.Y.; Crankshaw, J.; Crankshaw, D.J.; Garfield, R.E.; Daniel, E.E. Am. J. Physiol. 1980, 239, C66-C74.
32. Grover, A.K.; Kwan, C.Y.; Daniel, E.E. Am. J. Physiol. (submitted).
33. Daniel, E.E.; Grover, A.K.; Kwan, C.Y. Symp. on The Calcium Switch on Vertebrate Smooth Muscle, FASEB Mtg. 1981. Fed. Proc. (accepted for publication).
34. Daniel, E.E., Grover, A.K.; Kwan, C.Y. Pharmacological Congress, Tokyo, Japan, June 1981.
35. Matlib, M.A.; Crankshaw, J.; Garfield, R.E.; Crankshaw, D.J.; Kwan, C.Y.; Branda, L.A.; Daniel, E.E. J. Biol. Chem. 1979, 254, 1834-1840.
36. Kwan, C.Y., Garfield, R.E.; Daniel, E.E. J. Mol. Cell. ardiol. 1979, 11, 639-659.
37. Grover, A.K.; Kannan, M.S.; Daniel, E.E. Cell Calcium 1980, 1, 135-146.
38. Kwan, C.Y.; Lee, R.M.K.W.; Daniel, E.E. Blood Vessels 1981, 18, 171-186.
39. Sakai, Y.; McLean, J.; Grover, A.K.; Garfield, R.E.; Fox, J.E.T.; Daniel, E.E. Can. J. Physiol. Pharmacol. 1981, 59, 1260-1267.
40. Kwan, C.Y.; Sakai, Y.; Grover, A.K.; Lee, R.M.K.W. Mol. Physiol. 1981 (submitted).
41. Kwan, C.Y.; Grover, A.K.; Lee, R.M.K.W.; Daniel, E.E. FASEB Mtg. 1981, Fed. Proc. 40, 551.
42. Osa, T.; Fujino, T. Jap. J. Physiol. 1978, 28, 197-209.
43. Cheung, D.W., Daniel, E.E. J. Physiol. (Lond.) 1980, 309, 13-27.
44. Grover, A.K.; Kwan, C.Y.; Garfield, R.E.; McLean, J.; Fox, J.E.T.; Daniel, E.E. Can. J. Physiol. Pharmacol. 1980, 58 (9) 1102-1113.
45. Wolowyk, M.W.; Kidwai, A.M.; Daniel, E.E. Can. J. Biochem. 1971, 49, 376-384.
46. Grover, A.K.; Kwan, C.Y.; Crankshaw, J.; Daniel, E.E. Can. J. Physiol. Pharmacol. 1981, 59, 1128-1133.
47. Grover, A.K.; Crankshaw, J.; Garfield, R.E.; Daniel, E.E. Can. J. Physiol. Pharmacol. 1980, 58, 1202-1211.
48. Grover, A.K.; Kwan, C.Y.; Daniel, E.E. Am. J. Physiol. (Cell Biology) 1981, 240, C175-C182.
49. Van Breemen, C.; Aaronson, P.; Loutzenhiser, R. Pharmacol. Rev. 1979, 30, 167-208.

50. Droogmans, G.; Casteels, R. J. Gen. Physiol. 1979, 74, 57-70.
51. Brading, A.J.; Burnett, M.; Sneddon, P. J. Physiol. (Lond.) 1980, 306, 411-429.
52. Somlyo, A.P.; Somlyo, A.V.; Shuman, H.; Sloane, B.; Scarpa, A. Ann. N.Y. Acad. Sci. 1978, 307, 523-544.
53. Somlyo, A.P.; Somlyo, A.V.; Shuman, H. J. Cell Biol. 1979, 81, 316-335.
54. Vallieres, J.; Scarpa, A.; Somlyo, A.P. Arch. Biochem. Biophys. 1975, 170, 659-669.
55. Janis, R.A.; Crankshaw, J.; Daniel, E.E. Am. J. Physiol. 1977, 232, C50-C58.
56. Batra, S. Biochim. Biophys. Acta 1973, 305, 428-432.
57. Somlyo, A.:.; Somlyo, A.V.; Shuman, H.; Garfield, R.E. "Ionic actions on vascular smooth muscle with special regard to blood vessels". Ed. E. Betz. Springer Verlag, N.Y. 1976; pp 17-20.
58. Carsten, M.E.; Miller, J.D. Arch. Biochem. Biophys. 1980, 204, 404-412.

RECEIVED June 14, 1982.

5

Calmodulin: An Intracellular Ca^{2+} Receptor Protein

CHARLES O. BROSTROM, MARGARET A. BROSTROM, and DONALD J. WOLFF

University of Medicine and Dentistry of New Jersey, Rutgers Medical School, Department of Pharmacology, Piscataway, NJ 08854

Calmodulin is a Ca^{2+}-binding protein which has a broad, if not ubiquitous, distribution in the tissues of eukaryotes. The binding of Ca^{2+} occurs with high affinity (in the submicromolar to micromolar range) and results in changes of calmodulin conformation. Calmodulin physically interacts with and confers reversible, Ca^{2+}-dependent activation upon a number of enzymes which catalyze rate limiting reactions in various metabolic processes. Included among these enzymes are forms of cyclic nucleotide phosphodiesterase, adenylate cyclase, myosin light chain kinase, phosphorylase kinase, glycogen synthase kinase, NAD^+ kinase, and (Ca^{2+} + Mg^{2+}) ATPase. Various antipsychotic drugs, such as the phenothiazines, bind to the $Ca^{2+} \cdot$ calmodulin complex blocking the activation of these enzymes.

Calcium ion is widely recognized as a major regulator of intracellular metabolism in eukaryotes. The cation is frequently involved as a coupling factor linking diverse humoral stimuli to resultant cellular responses. Calmodulin appears to function as a major, if not the predominant, intracellular receptor for Ca^{2+}, coupling changes in intracellular free Ca^{2+} concentrations to subsequent cellular responses.

The evolution of primeval life forms into complex multicellular organisms required the concomitant development of effective intercellular communication systems for coordinating essential functions. In higher animals intercellular signaling occurs rapidly via the activation of electrochemical transmission through various neural networks or, more slowly, through the release of humoral substances into the circulation. By virtue of its regulatory actions on cellular processes, calcium ion functions

0097-6156/82/0201-0089$06.00/0

prominently in these signaling systems both in the generation and reception of messages. The elevation of intracellular free Ca^{2+} by a variety of stimuli is well-established to increase the activities of various enzyme systems involved in specialized cellular processes (1,2). Included are such processes as cell motility, muscle contraction, and exocytotic secretion from exocrine, endocrine, and nerve cells. The mechanisms by which various biochemical signals are selectively received and translated by the cell plasmalemma into increased concentrations of intracellular free Ca^{2+} are incompletely understood. It does appear likely, however, that the increased rate of phospholipid turnover in the plasmalemma observed in response to the binding of many hormones may prove to be involved in the creation of or opening of Ca^{2+} channels (3). Calcium ions would then proceed down a concentration gradient into the resting cell ($\simeq 10^{-7}$ M) from the interstitial fluid (1 mM) triggering subsequent responses. Alternatively, other systems, such as skeletal muscle, may predominantly involve release of internally sequestered Ca^{2+}. One of the more interesting features of intercellular communication is that a multitude of humoral signals (first messengers) are translated by the plasmalemma into changes in concentration of a restricted number of common intracellular denominators (second messengers). At present Ca^{2+} and adenosine 3',5'-monophosphate (cAMP) are the best established putative second messenger substances, although other candidates, such as guanosine 3',5'-monophosphate (cGMP), have also been suggested.

The translation of Ca^{2+}-dependent signals into biological responses should predictably involve intracellular Ca^{2+} receptors capable of rapidly and reversibly binding the cation in the range of concentrations believed to occur physiologically. Free Ca^{2+} concentrations are believed to vary from approximately 10^{-7} M in resting cells to perhaps 10^{-5} M in maximally stimulated cells. Various Ca^{2+}-binding proteins which bind Ca^{2+} in this range of concentrations have been isolated and characterized. These proteins include the two forms of troponin C from skeletal and cardiac muscle, parvalbumin, a vitamin D inducible protein, calsequestrin, and calmodulin. Calsequestrin appears to function exclusively in Ca^{2+} accumulation by endoplasmic reticulum. With the exception of calmodulin, the other proteins are of limited distribution and function in tissue specific responses (Table I).

Calmodulin, by contrast, is distributed throughout most, if not all, eukaryotic cells from both animal and plant sources. It has not been reported to exist in bacteria. Calmodulin varies in concentration from tissue to tissue with mammalian brain (4) and testis (5) and the electroplax of *Electrophorus electricus* (6) possessing particularly high content. While the protein has been found to be predominantly cytoplasmic in subcellular fractionation studies, substantial amounts are particulate-associated as well. Binding of calmodulin to particulate fractions is increased by Ca^{2+}, appears to occur at specific sites (7,8), and is saturable, reversible, and temperature and trypsin sensitive. Calmodulin has been found by indirect immunofluorescence techniques to be distributed generally throughout the cytoplasm of interphase proliferating cells in tissue culture (9,10). At mitosis the protein was found to associate with the mitotic apparatus. Calmodulin has been reported to associate with cytoplasm, nucleus, plasma membrane, and glycogen particles in rat liver tissue slices (11). Related distributions were found for adrenal and skeletal muscle slices.

The most interesting property of calmodulin is that it confers reversible, Ca^{2+}-dependent activation upon a series of enzymes (Table II). By virtue of physically interacting with a group of otherwise unrelated enzymes and shifting their respective conformations to active species, calmodulin serves in the capacity of a multifunctional calcium receptor protein. The versatility of calmodulin as an activator of diverse enzymes may be unique among non-catalytic regulatory proteins. It is this feature which has generated the remarkable current interest in the biochemistry and pharmacology of the molecule. Subsequent text in this article will be concerned with briefly detailing the properties, functions, and pharmacology of calmodulin with the emphasis on some of the most current advances. The text is conceived as an extension and updating of a recent review summarizing information through mid-1980 (12). For additional information and alternate viewpoints the reader is referred to a number of other recent reviews (13-18).

Physical and Chemical Properties of Calmodulin

Calmodulin has been purified to homogeneity from a variety of animal sources and characterized extensively. The protein is remarkably similar in terms of amino acid composition and tryptic peptide mapping patterns from one source to another (19). Comparisons of complete sequencing data for calmodulin from bovine brain (148 amino acid residues) (20) with partial sequencing data from bovine uterus (21), rat testis (22), and the coelenterate, *Renilla reniformis* (23) show only minor differences in primary structure. These differences almost exclusively involve conservative amino acid substitutions or changes in amidation states. The nucleotide sequence of the calmodulin structural gene also appears to be conserved. In a preliminary study (24), a cloned calmodulin structural gene corresponding to amino acid residues 93-148 from *Electrophorus electricus* has been found to hybridize with total genomic DNA preparations from electric eel kidney, hen liver, human placenta, and wheat germ. Discrete hybridization bands were found for all DNA preparations except that of wheat germ, which hybridized in a diffuse manner.

Based on a sequence of 148 amino acid residues, calmodulin has a calculated molecular mass of 16,800, which lies about midway in a series of reported values determined by sedimentation

Table I

Distribution and Putative Functions of Selected High Affinity Calciproteins

Protein	Distribution	Proposed Functions
calsequesterin	muscle,brain	intracellar Ca^{2+} storage
vitamin D inducible	intestine,kidney	Ca^{2+} absorption, reabsorption
troponin C	skeletal and cardiac muscle	regulation of contraction
parvalbumin	skeletal muscle	unknown
calmodulin	ubiquitous	multi-functional

Table II

Putative Calmodulin-Dependent Functions and Enzymes

Functions	Enzymes	Reviews or other citations
cyclic nucleotide metabolism	cyclic nucleotide phosphodiesterase	13,14,17
	adenylate cyclase	13,53
	guanylate cyclase (*Tetrahymena*)	52
glycogen metabolism	phosphorylase *b* kinase	48,54
	glycogen synthase kinase	55
smooth muscle contraction	myosin light chain kinase	13,17,18,56
NAD^+ conversion to $NADP^+$	NAD^+ kinase	57
Ca^{2+} transport	$(Ca^{2+} + Mg^{2+})$ATPase	13,17,18,58

equilibrium and sodium dodecylsulfate analytical gel electrophoresis methodology. The protein appears to exist primarily in solution in the monomeric form. Calmodulin from animal sources possesses one histidine, no tryptophan or cysteine, and one trimethyllysine residue. A low content of aromatic amino acids with a preponderance of phenylalanine to tyrosine generates the low molar extinction coefficient ($\epsilon_{276}^{1\%} = 1.8$). A predominance of acidic to basic amino acid residues yields a low isoelectric point (3.9) and the protein binds strongly to histones and other basic peptides (25-27).

The enzymatic methylation of calmodulin in rat brain cytosolic preparations has recently been investigated (28). Methyl group transfer from S-adenosylmethionine occurred principally into ϵ-trimethyllysine residues and was inhibited by EGTA. Methylation was stimulated by divalent cations, with Mn^{2+} providing the highest rate. It is unclear, however, whether the effect of the cation was enzyme or substrate directed. While the trimethyllysine residue of calmodulin has received considerable attention because of its rarity in protein structures, the residue is not apparently required for calmodulin to activate enzymes (29).

The amino acid sequence of calmodulin is arranged in four domains possessing substantial degrees of homology. Starting from the amino terminal end, the extent of homology is greatest between the first and third domains and between the second and fourth. One Ca^{2+} binding site is believed to reside in each domain. It has been suggested that these homologies may have originated from the occurrence of genetic redundancy in the genome for calmodulin in early ancestral cells (20).

As discussed previously (12), there is general agreement that calmodulin binds four Ca^{2+} per molecule. The values reported for the respective binding affinities, however, have been discrepant, apparently because determinations were conducted under varying experimental conditions. Most frequently two types of binding sites have been reported, based on data from equilibrium dialysis, with (depending on the report) one class of sites binding two or three Ca^{2+} with K_d values ranging from 0.2 to 4 μM and the other class binding one to three Ca^{2+} with K_d values ranging from 1 to 800 μM. There is also good evidence that other divalent cations compete for these sites raising the apparent K_d for Ca^{2+}. Such competition is apparently not observed in binding studies conducted with EGTA buffers (5,30). A reappraisal of Ca^{2+} binding to calmodulin utilizing equilibrium dialysis at 100 μM KCl has recently been reported (31). A series of K_d values ranging from 3 to 20 μM were calculated for the binding of Ca^{2+} with some indication of positive co-operativity (Hill co-efficient 1.33). Binding at the first site apparently induced conformational changes in the protein facilitating binding of Ca^{2+} at the second but not at the third or fourth sites. Conformational changes affecting the environment of the aromatic amino acid residues of calmodulin as measured by U.V. absorption and near U.V.-circular dichroic spectroscopy were largely completed upon the binding of 2 mol Ca^{2+}/mol protein, in agreement with previous findings regarding spectral (5,32-34) and NMR (35) changes occurring as a function of Ca^{2+} concentration.

It has been suggested that tyrosine fluorescence measurements conducted in conjunction with terbium binding to calmodulin allow the determination of the sequences of ion binding to the four domains (36). Through comparison of mammalian calmodulin, which has a tyrosyl residue at position 99 (domain III) and at position 138 (domain IV), with invertebrate calmodulin, which has only one tyrosyl residue (analogous to the mammalian position 138), the sequence of binding by this technique was reported to occur first at domains I and II (no tyrosyl residues), then to III, and then to IV.

A very recent binding study (37) appears to resolve some of the discrepancies in the earlier literature. Intrinsic binding constants for K^+, Mg^{2+}, and Ca^{2+} for each of the four cation binding sites were derived from flow dialysis data in conjunction with mathematical modeling. These cations bound competitively to the same sites. Binding constants determined for the four sites in the absence of competing cations ranged from 0.13 to 1.3 μM for Ca^{2+} and 1.5 to 11 mM for K^+. Constants for Mg^{2+} at 20 mM K^+ ranged from 0.4 to 1.5 mM for the four sites. These constants increased at 200 mM K^+ to range from 2 to 5 mM for three sites, the fourth being undetermined. From an extensive mathematical treatment, it appeared that each site had a somewhat different affinity for a given cationic species; however, no two cationic species had the same rank order of affinities. The rank order of affinities for Ca^{2+} appeared to support the earlier order of addition established for terbium binding (36). In the absence of competing cations or at 20 mM K^+, Ca^{2+} bound to three sites with similar, high affinities and at a fourth site with poorer affinity, as had been observed previously (33). The apparent positive co-operativity reported earlier (31) in the binding of Ca^{2+} to the first two sites of calmodulin arose in a complex manner involving the displacement of K^+.

Spectral changes accompanying the binding of Ca^{2+} are usually interpreted as reflecting an increase in helical content (33,38). While substantial spectral changes are reported in the literature, most measurements have involved comparison of the Ca^{2+}-saturated form with the inactive, Ca^{2+}-free form. Similar but somewhat less extensive conformational changes occur in calmodulin exposed to Mg^{2+}, without providing a complex capable of activating calmodulin dependent enzymes (33). It is likely that the conformational changes occurring in calmodulin allowing formation of an activating protein species *in vivo* involve displacement of Mg^{2+} and K^+ by Ca^{2+} and are probably rather subtle. In this regard it should be noted that the binding of 2 mol Ca^{2+}/mol calmodulin, which is sufficient to generate most of the observed spectral changes, is apparently inadequate to provide activation of the calmodulin-dependent phosphodiesterase (31). In addition to spectral changes, Ca^{2+} alters the sensitivity of the protein to proteolysis (39) and chemical modification (38, 40-41) and increases the surface hydrophobicity of the protein (42), all of which are indicative of the occurrence of conformational changes.

Native calmodulin has in general proven to be a poor antigen for the reproducible preparation of antisera of useful titer and of high affinity, although there have been some reported successes (43). Most of the antibodies which have been prepared against either native or derivitized calmodulin have not differentiated between the Ca^{2+}-free and Ca^{2+}-saturated conformation. A recent study has compared the efficacy of various calmodulin derivatives in producing reproducible and high titer antisera (44). Calmodulin preparations injected as either the native protein or injected following treatment with sodium dodecylsulfate, coupling to hemocyanin with carbodiimide, derivitized with dinitrophenol, or adsorbed to alumina did not elicit reproducible production of high titer antisera. Performic acid oxidation of the methionyl amino acid residues to methionyl sulfone residues provided rapid and reproducible production of anti-calmodulin sera of high specificity, affinity, and titer for both the derivative and native calmodulin. These antisera did not, however, differentiate between the Ca^{2+}-free and Ca^{2+}-saturated conformations. A major immunoreactive site was proposed to reside in the 18 residues at the carboxyterminal end of calmodulin on the basis of the reactivity of peptides prepared by trypsin and cyanogen bromide cleavage.

An azido-^{125}I-calmodulin derivative suitable for labeling calmodulin binding proteins by photoaffinity has recently been described (45). Azido-calmodulin retained the ability to interact with and activate the calmodulin-dependent phosphodiesterase in a Ca^{2+}-dependent manner. Upon photolysis calmodulin-binding subunits of the enzyme formed 1:1 crosslinked complexes with the derivative. Crosslinked products were not obtained with incubations performed with EGTA or in large excess of unmodified calmodulin. Following photoaffinity labeling the calmodulin-dependent ATPase from red cells was found to be irreversibly activated by the derivative.

Crystallization of calmodulin from rat testis (46) and bovine brain (47) has recently been reported. Somewhat different methods of crystallization were utilized in the two preparations, with crystallization apparently occurring in different modes. Testis calmodulin crystals grown from solutions of 2-methyl-2,4-pentanediol were described as being triclinic, space group P1, with one calmodulin molecule per unit cell, and diffracting with resolution beyond 2.5 A. Brain calmodulin crystals produced in solutions of polyethylene glycol were described as being of space group $P2_1$ with two monomers calmodulin per unit cell, and diffracting with resolution beyond 5 A.

Mechanism of Enzyme Activation by Calmodulin

Current understanding of the mechanism by which calmodulin induces enzyme activation has been discussed in some detail previously (12). All enzymes known to be activated by calmodulin (Table II) with the exception of phosphorylase b kinase (48), readily dissociate from the protein on anion exchange columns when chromatographed with Ca^{2+} chelators. Regulation of dissociable enzymes by calmodulin (C) is generally believed to occur through two sequential, fully reversible mass action expressions:

$$nCa^{2+} + C \rightleftharpoons \left(Ca^{2+}\right)_n C \qquad (1)$$

$$x\left(Ca^{2+}\right)_n C + \text{enzyme}_{(inact)} \rightleftharpoons [(Ca^{2+})_n C]_x \cdot \text{enzyme}_{(act)} \qquad (2)$$

In this model Ca^{2+} first binds to calmodulin producing a protein conformation capable of enzyme activation (Eqn. 1). This species then associates with the inactive form of the enzyme forming an activated ternary complex (Eqn. 2). Enzyme activation appears to result from a conformational change induced in the enzyme structure from the binding of X Ca^{2+}·calmodulin complexes. At present the number (n) of Ca^{2+} required for enzyme activation is in dispute. Indeed, it has been suggested that the number may vary from one enzyme species to another (36,37). At 1-3 mM Mg^{2+} an activating complex for the cyclic nucleotide phosphodiesterase with n equal to 3 (30,33) but not n equal to 2 (31) has been reported. Another report, however, derived an n equal to 4 for this enzyme (49). An n equal to 4 has also been reported for the myosin light chain kinase (50). These determinations are largely based upon complex analyses of kinetic data and involve various assumptions, some of which might not necessarily be irrefutable.

The value of X (Eqn. 2) appears to be variable depending on the number of binding subunits for a given enzyme. For example, the cyclic nucleotide phosphodiesterase is frequently described as being a dimeric molecule with each of the two monomers capable of binding one molecule of calmodulin. Similarly, the myosin light chain kinase is reported to bind one calmodulin molecule per subunit (51). Multiple calmodulin binding sites per monomer have not as yet been described for any enzyme with the possible exception of phosphorylase b kinase.

Dissociation of the ternary complex presumably could proceed via a direct reversal of Eqn. 2 or as a consequence of a dissociation of Ca^{2+} from the ternary complex followed by reversion of calmodulin to an inactive conformation and separation from the enzyme. Data are not currently available to distinguish whether one or both types of dissociation prevail. It is also not known whether the ternary complex has a higher affinity for Ca^{2+} than does free calmodulin. As discussed previously (12), the Ca^{2+} sensitivity for enzymes produced by this model would be a function of (a) the concentration of calmodulin in the system, (b) the concentration of cations competing for the Ca^{2+} binding sites, and (c) the relative affinity of the inactive enzyme for the Ca^{2+}-calmodulin complex. While the above model is consistent with the literature currently available regarding the activation of various enzymes by calmodulin, the possibility exists that it may ultimately prove to be overly simplified.

Regulatory Roles of Calmodulin

As emphasized in Table II, calmodulin confers Ca^{2+}-dependent activation upon a remarkable variety of enzymes, with others probably remaining to be discovered. Each of these enzymes appears to catalyze a rate limiting reaction in the tissues where it is found. For systems possessing these enzymes, calmodulin appears to function, in effect, as an intracellular Ca^{2+}-receptor protein which couples stimuli-provoked changes in free Ca^{2+} concentration to enzyme activation and the generation of cellular responses. While there is evidence that a multiplicity of cell functions may be potentially regulated in this manner (Table II), species and tissue variations exist. That is, a given process is not necessarily subject to calmodulin-dependent regulation in each cell type where it occurs. For example, the calmodulin-dependent forms of cyclic nucleotide phosphodiesterase and adenylate cyclase are of limited tissue distribution with the phosphodiesterase being the more widely distributed of the two (13). Also, although guanylate cyclase is a widely distributed activity, only *Tetrahymena* has been found to possess a calmodulin-dependent form (52).

Space does not permit a detailed review of the biochemistry of each of the enzymes activated by calmodulin, although an extensive and interesting literature is available. The salient features of some of the more recent literature pertaining to these enzymes is available elsewhere (12). More extensive reviews also exist for the older literature pertaining to most of these enzymes (see citations in Table II).

Pharmacologic Considerations

The possibilities for pharmacologic intervention in calmodulin-dependent processes would appear to be quite promising, although largely undeveloped at this time. Modes of intercession could include the development of (a) inhibitors of activation acting on the calmodulin molecule itself or on one or more of the activatable enzymes, (b) agents shifting the conformation of calmodulin to the active form or increasing the affinity of activatable enzymes for the $Ca^{2+}\cdot$calmodulin complex, and (c) drugs which substitute for calmodulin in the activation of one or more of these enzymes. As discussed previously (12), very few substances have been shown to substitute for calmodulin as alternate enzyme activators. For example, among related Ca^{2+} binding proteins even troponin C, which has a 77% amino acid sequence homology with calmodulin, does not appear to activate calmodulin-dependent enzymes. Good evidence, however, has been reported that a 11,500 Mr Ca^{2+} binding protein isolated from rat hepatoma will activate the calmodulin-dependent phosphodiesterase (59). Based on different tryptic peptide mapping patterns and a lack of immunocrossreactivity, the hepatoma protein appeared to be structurally different from calmodulin. Enzyme activation required approximately 10-fold more tumor protein than calmodulin on a molar basis, occurred at an equivalent Ca^{2+} concentration, and was reversed by Ca^{2+} chelation. The calmodulin-dependent phosphodiesterase (60) and ATPase (61) activities can be activated by various phospholipids independent of the presence of Ca^{2+} or calmodulin. These observations provide a basis for predicting that pharmacologic substitutes for calmodulin are potentially derivable. Pharmacologic agents have not as yet been developed which shift the apparent K_m for calmodulin of dependent enzymes. In this context, however, it should be noted that basic proteins (26-27) and several calmodulin-binding proteins of unknown function (62-63) have been reported to increase the apparent K_m for calmodulin of various enzymes. Also Mg^{2+} reportedly decreases the apparent K_m of the phosphodiesterase for calmodulin (64), and GTP and F^- decrease the apparent K_m of the adenylate cyclase for calmodulin (65).

Various psychoactive drugs are well-established to inhibit calmodulin-dependent enzymes, with the phenothiazine antipsychotics currently being among the most potent agents (12,13,66). Such inhibition is reversed by increasing the concentration of calmodulin in the incubation (60,65,67-68) but not the concentration of Ca^{2+}. These inhibitory actions arise as a consequence of the binding of 2 moles phenothiazine with high affinity (K_d 1-10 μM) to the Ca^{2+}·calmodulin complex (69). Binding of the phenothiazines to the protein is rapid and normally considered to be Ca^{2+}-dependent, although the drugs will also bind to calmodulin complexed with Sr^{2+}, Ni^{2+}, Co^{2+}, Zn^{2+} or Mn^{2+} but not with Mg^{2+} or Ba^{2+} (69). Binding is reversed by chelating agents (69) and is non-stereospecific with respect to other antipsychotic drugs which have stereoisomers (70,71). Phenothiazines are believed to bind to the active conformation of calmodulin preventing the protein from interacting with and activating calmodulin-dependent enzymes.

Although calmodulin has certain properties which might be predicted for a phenothiazine receptor protein, it is inappropriate to assume that all enzymes inhibited by these agents have a calmodulin requirement. The phenothiazines have lipid solubilities and detergent properties that provide a wide range of effects on membranes and membrane-associated processes (72,73) which are probably independent of calmodulin. Forms of adenylate cyclase found not to be activated by Ca^{2+} and calmodulin are inhibited by phenothiazines (74). For example the dopamine-sensitive adenylate cyclase from brain is exquisitely sensitive to inhibition by trifluoperazine ($K_i = 8 \times 10^{-9}$ M) (75). It is also of interest that the activation of the cyclic nucleotide phosphodiesterase by the non-calmodulin, hepatoma Ca^{2+} binding protein discussed above is inhibited by phenothiazine (59). Troponin C, which is not an activator of the phosphodiesterase, also binds phenothiazines (69).

It has been reported that high affinity, Ca^{2+}-dependent binding to calmodulin appears to be a general property of clinically effective antipsychotic agents and that this binding parallels the potency of these compounds to inhibit calmodulin-dependent phosphodiesterase activity (70). Antianxiety and antidepressant drugs bound more weakly and were less effective inhibitors. Other agents such as LSD, amphetamine, phenobarbital, morphine, and various biogenic amines that affect the central nervous system but are devoid of antipsychotic activity, neither bound nor inhibited. In general the affinity of binding to calmodulin qualitatively paralleled the potency of the agents as antipsychotics and their ability to produce extrapyramidal effects. Some lack of stereospecificity was found in this study. A later, more extensive investigation of stereospecificity found that both the clinically active and inactive isomers of a series of neuroleptics were equally effective inhibitors of the phosphodiesterase (71). The IC_{50} values for inhibition of the enzyme correlated closely with the octanol:H_2O partition coefficients of the drugs; and it was argued that the observed lack of stereospecificity and the relatively high concentration of drug required for enzyme inhibition effectively ruled out a calmodulin involvement in the therapeutic actions of the drugs. More recently the specificity of the bnding of phenothiazine to calmodulin has been examined for a series of chlorpromazine analogs differing in the position of chlorine substituent on the aromatic nucleus (76). Of these compounds only the 2-chloro analog had tranquilizer activity and antagonized dopamine-sensitive adenylate cyclase activity, but all possessed similar hydrophobicity, membrane actions, and surface activity. All of the chlorpromazine derivatives inhibited the calmodulin-dependent ATPase from red cells with similar potencies. The failure of calmodulin to discriminate among these compounds was taken as further evidence that calmodulin lacks a phenothiazine binding site with specificity similar to the receptors responsible for the antipsychotic effects of the drug observed clinically.

In addition to the psychoactive drugs, various agents with local anesthetic-like properties bind weakly in a Ca^{2+}-dependent manner (77). Another agent found to bind in the presence of Ca^{2+} and inhibit calmodulin-dependent enzymes is the sulfonamide, *N*-(6-aminohexyl)-5-chloro-1-naphthalene sulfonate (78). Also various classes of hydrophobic dyes, such as 9-anthroylcholine, bind to calmodulin with Ca^{2+} and prevent enzyme activation (42). From this data it appears reasonably clear that the active conformation of calmodulin possesses two or more hydrophobic pockets capable of binding a wide variety of lipophilic organic molecules. While the binding affinities generally tend to parallel such criteria as octanol:H_2O solubilities, it is likely that binding may also involve various steric considerations and interactions with functional groups on the molecules. Presumably the high affinities of binding observed for certain of the phenothiazines arise as a consequence of a particularly fortuitous balance of these variables. The hydrophobic surface pockets exposed by Ca^{2+} are apparently critically involved in the interaction of the calmodulin with and activation of dependent enzymes. Presumably these enzymes possess matching hydrophobic regions which interact with these sites. As noted above the phosphodiesterase (60) and ATPase (61) are activated without Ca^{2+} or calmodulin by phospholipids, possibly acting at such sites. Interestingly, such activation of the phosphodiesterase is reversed by phenothiazines (60). It is conceivable that phenothiazines bind not only to calmodulin but also to matching sites on the phosphodiesterase and the other activatable enzymes.

Acknowledgement

Preparation of the manuscript was supported by U. S. Public Health Service grants NS 11340, NS 11252, and AM 28099.

Literature Cited

1. Rasmussen, H.; Goodman, D. B. P. Pharmacol. Rev. 1977, 57, 421.
2. Berridge, M. J. Adv. Cyclic Nucleo. Res. 1975, 6, 1.
3. Michell, R. H. TIBS 1978, 4, 128.
4. Watterson, D. M.; Harrelson, W. G.; Keller, P. M.; Sharief, F.; Vanaman, T. C. J. Biol. Chem. 1976, 251, 4501.
5. Dedman, J. R.; Potter, J. D.; Jackson, R. L.; Johnson, J. D.; Means, A. R. J. Biol. Chem. 1977, 252, 8415.
6. Childers, S. R.; Siegel, F. L. Biochim. Biophys. Acta 1975, 405, 99.
7. Kakiuchi, S.; Yamazaki, R.; Teshima, Y.; Uenishi, K.; Yasuda, S.; Kashiba, A.; Sobue, K.; Oshima, M.; Nakajima, T. Adv. Cyclic Nucleo. Res. 1978, 9, 253.
8. Vandermeers, A.; Robberecht, P.; Vandermeers-Piret, M.; Rathe, J.; Christopher, J. Biochem. Biophys. Res. Commun. 1978, 84, 1076.
9. Welsh, M. J.; Dedman, J. R.; Brinkley, B. R.; Means, A. R. Proc. Natl. Acad. Sci. U.S.A. 1978, 75, 1867.
10. Andersen, B.; Osborn, M.; Weber, K. Cytobiologie 1978, 17, 354.

11. Harper, J. F.; Cheung, W. Y.; Wallace, R. W.; Huang, H. L.; Levine, S. N.; Steiner, A. L. Proc. Natl. Acad. Sci. U.S.A. 1980, 77, 366.
12. Brostrom, C. O.; Wolff, D. J. Biochem. Pharmacol. 1981, 30, 1395.
13. Wolff, D. J.; Brostrom, C. O. Adv. Cyclic Nucleo. Res. 1979, 11, 27.
14. Wang, J. H.; Waisman, D. M. Curr. Topics Cell Regulat. 1979, 15, 47.
15. Cheung, W. Y. Science 1980, 207, 19.
16. Means, A. R.; Dedman, J. R. Nature (Lond.) 1980, 285, 73.
17. Klee, C. B.; Crouch, T. H.; Richman, P. G. Ann. Rev. Biochem. 1980, 49, 489.
18. Walsh, M. P.; LePeuch, C. J.; Vallet, B.; Cavadore, J. C.; Demaille, J. G. J. Mol. Cell. Cardiol. 1980, 12, 1091.
19. Vanaman, T. C.: Sharief, F.; Awramik, J. L.; Mandel, P. A.; Watterson, D. "Contractile Systems in Non-Muscle Systems"; Perry, S. V.; Margreth, A.; Adelstein, R.S., Eds; North Holland-Elsevier: Amsterdam, 1976; pp. 165-176.
20. Watterson, D. M.; Sharief, F.; Vanaman, T. C. J. Biol. Chem. 1980, 255, 962.
21. Grand, R. J. A.; Perry, S. V. Fed. Eur. Biochem. Soc. Lett. 1978, 92, 137.
22. Dedman, J. R.; Jackson, R. L.; Schreiber, W. E.; Means, A. R. J. Biol. Chem. 1978, 253, 343.
23. Goodman, M.; Pechere, J. F.; Haiech, J.; Demaille, J. G. J. Mol. Evol. 1979, 13, 331.
24. Munjaal, R. P.; Chandra, T.; Woo, S. L. C.; Dedman, J. R.; Means, A. R. Proc. Natl. Acad. Sci. U.S.A. 1981, 78, 2330.
25. Khandelwal, R. L.; Kotello, M. C.; Wang, J. H. Arch. Biochem. Biophys. 1980, 203, 244.
26. Itano, T.; Itano, R.; Penniston, J. T. Biochem. J. 1980, 189, 455.
27. Wolff, D. J.; Ross, J. M.; Thompson, P. N.; Brostrom, M. A.; Brostrom, C. O. J. Biol. Chem. 1981, 256, 1846.
28. Sitaramayya, A.; Wright, L. S.; Siegel, F. L. J. Biol. Chem. 1980, 255, 8894.
29. Molla, A.; Kilhoffer, M. C.; Ferraz, C.; Audemard, E.; Walsh, M. P.; Demaille, J. G. J. Biol. Chem. 1981, 256, 15.
30. Cox, J. A.; Malnoe, A.; Stein, E. A. J. Biol. Chem. 1981, 256, 3218.
31. Crouch, T. H.; Klee, C. B. Biochemistry 1980, 19, 3692.
32. Klee, C. B. Biochemistry 1977, 16, 1017.
33. Wolff, D. J.; Poirier, P. G.; Brostrom, C. O.; Brostrom, M. A. J. Biol. Chem. 1977, 252, 4108.
34. Yagi, K.; Yazawa, M.; Kakiuchi, S.; Oshima, M.; Uenishi, K. J. Biol. Chem. 1978, 253, 1338.
35. Seamon, K. B. Biochemistry 1980, 19, 207.
36. Kilhoffer, M. C.; Demaille, J. G.; Gerard, D. Fed. Eur. Biochem. Soc. Lett. 1980, 116, 269.

37. Haiech, J.; Klee, C. B.; Demaille, J. G. Biochemistry 1981, 20, 3890.
38. Liu, Y. P.; Cheung, W. Y. J. Biol. Chem. 1976, 251, 4193.
39. Ho, H. C.; Desai, R.; Wang, J. H. Fed. Eur. Biochem. Soc. Lett. 1975, 50, 374.
40. Walsh, M.; Stevens, F. C. Biochemistry 1977, 16, 2742.
41. Richman, P. G.; Klee, C. B. Biochemistry 1978, 17, 928.
42. LaPorte, D. C.; Wierman, B. M.; Storm, D. R. Biochemistry 1980, 19, 3814.
43. Chafouleas, J. G.; Dedman, J. R.; Means, A. R. J. Biol. Chem. 1979, 254, 10.
44. Eldik, L. J.; Watterson, D. M. J. Biol. Chem. 1981, 256, 4205.
45. Andreasen, T. J.; Keller, C. H.; LaPorte, D. C.; Edelman, A. M.; Storm, D. M. Proc. Natl. Acad. Sci. U.S.A. 1981, 78, 2782.
46. Cook, W. J.; Dedman, J. R.; Means, A. R.; Bugg, C. E. J. Biol. Chem. 1980, 255, 8152.
47. Kretsinger, R. H.; Rudnick, S. E.; Sneden, D. A.; Schatz, V. B. J. Biol. Chem. 1980, 255, 8154.
48. Cohen, P.; Burchell, A.; Foulkes, J. G.; Cohen, P. T. W.; Vanaman, T. C.; Nairn, A. C. Fed. Eur. Biochem. Soc. Lett. 1978, 92, 287.
49. Huang, C. Y.; Chau, V.; Chock, P. B.; Wang, J. H.; Sharma, R. K. Proc. Natl. Acad. Sci. U.S.A. 1981, 78, 871.
50. Blumenthal, D. K.; Stull, J. T. Biochemistry 1980, 19, 5608.
51. Conti, M. A.; Adelstein, R. S. J. Biol. Chem. 1981, 256, 3178.
52. Kakiuchi, S.; Sobue, K.; Yamazaki, R.; Nagao, S.; Umeki, S.; Nozawa, Y.; Yazawa, M.; Yagi, K. J. Biol. Chem. 1981, 256, 19.
53. Bradham, L. S.; Cheung, W. Y. Calcium and Cell Function 1980, 1, 109.
54. Walsh, K. X.; Milliken, D. M.; Schlender, K. K.; Reimann, E. M. J. Biol. Chem. 1980, 255, 5036.
55. Payne, M. E.; Soderling, T. R. J. Biol. Chem. 1980, 255, 8054.
56. Adelstein, R. S. Ann. Rev. Biochem. 1980, 49, 921.
57. Anderson, J. M.; Charbonneau, H.; Jones, H. P.; McCann, R. O.; Cormier, M. J. Biochemistry 1980, 19, 3113.
58. Roufogalis, B. D. Can. J. Physiol. Pharmacol. 1979, 57, 1331.
59. MacManus, J. P. Fed. Eur. Biochem. Soc. Lett. 1981, 126, 245.
60. Wolff, D. J.; Brostrom, C. O. Arch. Biochem. Biophys. 1976, 173, 720.
61. Niggli, V.; Adunyah, E. S.; Penniston, J. T.; Carafoli, E. J. Biol. Chem. 1981, 256, 395.
62. Wang, J. H.; Desai, R. J. Biol. Chem. 1977, 252, 4175.
63. Wallace, R. W.; Tallant, E. A.; Cheung, W. Y. Biochemistry 1980, 19, 1831.

64. Brostrom, C. O.; Wolff, D. J. Arch. Biochem. Biophys. 1976, 172, 301.
65. Brostrom, M. A.; Brostrom, C. O.; Wolff, D. J. Arch. Biochem. Biophys. 1978, 191, 341.
66. Weiss, B.; Prozialeck, W.; Cimino, M. Adv. Cyclic Nucleo. Res. 1980, 12, 213.
67. Hidaka, H.; Yamaki, T.; Tatsuka, T.; Asano, M. Mol. Pharmacol. 1979, 15, 49.
68. Raess, B. U.; Vincenzi, F. F. Mol. Pharmacol. 1980, 18, 253.
69. Levin, R. M.; Weiss, B. Mol. Pharmacol. 1977, 13, 690.
70. Levin, R. M.; Weiss, B. J. Pharmacol. Exptl. Therap. 1979, 208, 454.
71. Norman, J. A.; Drummond, A. H.; Moser, P. Mol. Pharmacol. 1979, 16, 1089.
72. Seeman, P. Pharmacol. Rev. 1972, 24, 583.
73. Seeman, P.; Staiman, A.; Chau-Wong, M. J. Pharmacol. Exptl. Therap. 1974, 190, 123.
74. Wolff, J.; Jones, A. B. Proc. Natl. Acad. Sci. U.S.A. 1970, 65, 454.
75. Clement-Cormier, Y. C.; Kebabian, J. W.; Petzold, G. L.; Greengard, P. Proc. Natl. Acad. Sci. U.S.A. 1974, 71, 1113.
76. Roufogalis, B. D. Biochem. Biophys. Res. Commun. 1981, 98, 607.
77. Volpi, M.; Sha'afi, R. I.; Epstein, P. M.; Andrenyak, D. M.; Feinstein, M. B. Proc. Natl. Acad. Sci. U.S.A. 1981, 78, 795.
78. Hidaka, H. T.; Naka, Y. M.; Tanaka, T.; Hayashi, H.; Kobayashi, R. Mol. Pharmacol. 1980, 17, 66.

RECEIVED June 8, 1982.

6

Pharmacology of Intracellular Calcium Antagonistic Methylenedioxyindenes

RALF G. RAHWAN and DONALD T. WITIAK

Ohio State University, College of Pharmacy, Division of Pharmacology and Division of Medicinal and Natural Products Chemistry, Columbus, OH 43210

Extensive pharmacological evidence supports the contention that 2-n-propyl- and 2-n-butyl-3-dimethylamino-5,6-methylenedioxyindenes (pr-MDI and bu-MDI, respectively) are calcium antagonists with a predominantly intracellular site of action. Support for this mechanism derives from the following findings: (1) their ability to interfere with barium-induced nonvascular smooth muscle contraction; (2) the reversibility of their vascular and nonvascular smooth muscle relaxant properties by increasing extracellular calcium; (3) their ability to inhibit calcium-dependent (but not calcium-independent) evoked adrenomedullary catecholamine secretion without interfering with cellular calcium uptake; (4) their ability to reduce the quantity of calcium released from sarcoplasmic reticulum upon stimulation as evidenced by depression of activation heat in skeletal muscle; (5) their inhibitory effect on caffeine-induced contracture of skeletal muscle in the presence and in the absence of extracellular calcium; (6) their inhibitory effect on thrombin-induced platelet secretion; (7) their binding characteristics to cardiac troponin-C and to brain calmodulin with resultant inhibition of calcium-calmodulin-dependent processes; (8) their inhibitory effect on swelling and uncoupling of oxidative-phosphorylation induced by inorganic phosphates in isolated rabbit heart mitochondria; (9) their ability to inhibit the contractile effects of U44069 on the rat aorta in a calcium-free medium; (10) their ability to block both phases of the contractile effect of norepinephrine on the isolated rat aorta in a calcium-free medium; (11) their inability to block myocardial membrane slow calcium channels or

0097-6156/82/0201-0105$06.00/0

other presumptive membrane calcium channels; (12) their ability to uncouple excitation-contraction coupling in superfused canine papillary muscle preparations at concentrations which do not reduce action potential characteristics; and (13) the relative inactivity in-vitro of the quaternary ammonium derivatives of the MDIs on inotropy of the electrically-driven guinea pig left atrium and on potassium-induced and norepinephrine-induced contractions of the isolated rat aortic strip. Pr-MDI and bu-MDI exhibit negative inotropic actions in isolated perfused intact rabbit hearts and in isolated guinea pig atrial preparations; coronary dilating properties in rabbit perfused hearts; hypotensive action in dogs; and antiarrhythmic activity in three animal species (dog, rat and mouse) using five models of experimentally-induced arrhythmias involving calcium, ouabain, aconitine, methachoine, and chloroform-anoxia. The quaternary derivative of bu-MDI demonstrated potent antiarrhythmic activity in-vivo, possibly due to metabolic activation.

Calcium antagonism is a property attributed to a wide spectrum of pharmacological agents (1). These agents are of considerable therapeutic value, particularly (but not exclusively) in the management of cardiovascular diseases (1-6). Calcium antagonistic drugs have been broadly classified into two major classes (1,7,8,9): those agents acting on the cell membrane to block calcium influx through the slow inward calcium channels, and those acting intracellularly to block the action or mobilization of intracellular calcium or to enhance the sequestration or efflux of this cation. With respect to the membrane calcium channel blockers, mounting evidence supports an additional intracellular calcium antagonistic effect of these agents (10-19), as well as other pharmacological actions exhibited by various agents including inhibition of the fast inward sodium channels and blockade of adrenergic receptors (8).

A series of 2-substituted 3-dimethylamino-5,6-methylenedioxyindenes (MDIs) were developed in our laboratories (20) as intermediates in the synthesis of potential prostaglandin antagonistic indanpropionic acids (21,22). Pharmacological evaluation of these intermediate MDIs demonstrated their calcium antagonistic properties (7,8,9). The following discussion is an update of a recent review on the pharmacology of the MDIs (8).

Basic Pharmacological Studies with the MDIs

The pr-MDI and bu-MDI ($5x10^{-5}$ - 10^{-4}M) blocked the spasmogenic action on the estrogenized rat uterus of $PGF_{2\alpha}$, PGE_2, oxytocin, barium, acetylcholine, and ergonovine in a concentration-dependent and reversible manner (23). Furthermore, the MDIs blocked the contractile effect of histamine on the isolated guinea pig ileum, and of acetylcholine on the isolated rat ileum (23). The antagonism by the MDIs of the spasmogenic action of acetylcholine (which utilizes extracellular calcium) and barium (which utilizes intracellular calcium) could be reversed by increasing the extracellular calcium concentration (23), indicating that the MDIs were interfering with excitation-contraction coupling by acting as calcium antagonists with a probably intracellular site of action.

To further characterize the calcium-antagonistic mechanism of action of the MDIs, we utilized the isolated perfused bovine adrenal medulla as a model for stimulus-secretion coupling (24), since this preparation has been shown to share many common molecular features with excitation-contraction coupling in muscle (25,26,27). Adrenal catecholamine secretion induced by carbachol is known to be mediated by extracellular calcium (28) and not by intracellular calcium (29). On the other hand, acetaldehyde-induced adrenomedullary catecholamine secretion is independent of both extracellular calcium (30,31,32) and intracellular calcium (30). The pr-MDI and bu-MDI (10^{-8} - 10^{-4}M) reversibly blocked adrenomedullary catecholamine secretion evoked by carbachol, but did not affect acetaldehyde-induced catecholamine secretion (24). Furthermore, these MDIs did not affect ^{45}Ca uptake by adrenomedullary chromaffin cells (24). These findings indicated that the MDIs were interfering with stimulus-secretion coupling by acting as calcium antagonists with a probably intracellular site of action.

Studies on skeletal muscle also support an intracellular site of action of the MDIs. Thus, pr-MDI (10^{-4}M) significantly blocked caffeine-induced contractures of the rat diaphragm both in presence and in absence of extracellular calcium (33). Such caffeine-induced contractures are believed to be mediated by intracellular calcium mobilized from the sarcoplasmic reticulum or other intracellular calcium pool (34). Furthermore, bu-MDI (10^{-4}M) depresses activation heat in the frog sartorius muscle upon stimulation (35), indicating a reduction in the quantity of calcium released from the sarcoplasmic reticulum, since activation heat represents the energy liberated in association with calcium mobilization and sequestration in contracting muscle (36, 37).

U44069 is a stable analogue of prostaglandin H_2 which contracts the aorta by mobilizing intracellular calcium (38). Bu-MDI (10^{-4}M) inhibited the contractile effect of U44069 on the isolated rat aorta in a calcium free medium (39), confirming an intracellular site of action of the MDI. Under similar experimental conditions, even high concentrations of nifedipine (a calcium channel blocker) did not block the effects of U44069 (39).

We have recently reported that norepinephrine produces a biphasic contraction of the isolated rat aortic strip in calcium-free-EGTA medium by mobilizing two distinct and dissociable intracellular pools of calcium associated with the two dissociable phases of contraction (40). Bu-MDI (10^{-4}M) inhibits both phases of the norepinephrine response, while nifedipine has no effect on either response (41).

If the tertiary pr-MDI and bu-MDI gain access to the interior of the cell to exert their calcium antagonistic actions, it would be predicted that their quaternary ammonium analogues would be inactive due to their exclusion from the intracellular compartment as a result of their inability to cross biological membranes. As predicted, the tertiary, but not the quaternary, MDIs produce a negative inotropic effect on the isolated electrically-driven guinea pig left atrium (42) and inhibit potassium-induced and norepinephrine-induced contractions of the isolated rat aortic strip (43).

In order to ascertain the absence of a membrane blocking effect of the MDIs on the slow inward calcium channels, we investigated the effect of these agents on the positive inotropic concentration-response curves of calcium, ouabain, and isoproterenol on the isolated electrically-driven guinea pig left atrium (44). The positive inotropic action exerted by elevation of the extracellular calcium concentration appears to be due to a direct increased entry of this cation into myocardial cells through the slow inward calcium channels (45). Beta-adrenergic agonists such as isoproterenol stimulate cardiac excitation-contraction coupling by indirectly augmenting the entry of extracellular calcium through the slow inward calcium channels (46,47,48) subsequent to activation of adenylate cyclase and eventual cyclic-AMP-mediated phosphorylation of sarcolemmal proteins (48). On the other hand, most of the published evidence argues against an influence of digitalis glycosides on the slow inward calcium current (49-52) and favors a mechanism of increased calcium influx through other membrane channels resulting from an alteration of the movement of a sarcolemmal Na^{+}/Ca^{++} transport system secondary to an increase in intracellular sodium concentration (47). Although excess extracellular calcium, isoproterenol, and ouabain reversed the negative inotropic effect of pr-MDI, an analysis of the concentration-response relationships to all three positive inotropic agents in the presence and in the absence of pr-MDI demonstrated that this agent did not significantly inhibit the contractile effects of calcium, isoproterenol, or ouabain, at pr-MDI concentrations of 10^{-4}M or less which clearly exhibit intrinsic negative inotropic effects (44). From this it is concluded that the MDIs do not block the membrane slow inward calcium channels nor other presumptive membrane routes of calcium entry into myocardial cells (44). These findings differ clearly from those reported for the membrane slow calcium channel blockers, verapamil and D600, which exhibit competitive antagonism against calcium and noncompetitive antagonism against isoproterenol in analyses of concentration-response relationships (45).

Molecular Pharmacological Studies with the MDIs

In order to further characterize the intracellular mechanism of action of the MDIs, their interaction with intracellular calcium receptors was examined (53). It is now believed that many of the second messenger effects of calcium are mediated by the intracellular calcium binding proteins (receptors), calmodulin and troponin-C (54,55,56). Equilibrium dialysis showed that bu-MDI bound to calmodulin in a calcium-dependent manner at four equivalent sites of equal affinity (6.2×10^{-4}M), over the range of 10^{-7} to 10^{-5}M free calcium which would be expected to occur physiologically. One-half maximal binding occurred at a free calcium concentration of 7×10^{-6}M. The interaction between bu-MDI and calmodulin was competitive in nature, and bu-MDI was capable of inhibiting a number of calcium-calmodulin-dependent processes (53). Furthermore, bu-MDI was capable of binding to cardiac troponin-C (53). Such findings indicate that the MDIs may interfere with calcium-dependent processes through their interaction with intracellular calcium receptors. Prenylamine (47) and felodipine (19), two calcium-channel blockers, have also been shown to interact with calcium-calmodulin and inhibit certain calcium-calmodulin-dependent processes.

Recent evidence suggests that ischemia of myocardial tissue results in an increase in inorganic phosphate concentration which induces mitochondrial swelling and uncoupling of oxidative-phosphorylation processes by stimulating energy-dissipating intramitochondrial cycling of calcium (57). Mitochondrial swelling induced by inorganic phosphate could be inhibited by the tertiary MDIs as well as by the calcium-channel blockers verapamil and diltiazem, but not by divalent metal chelators (58). The net results of the protective effects of calcium antagonists on mitochondrial swelling are an increase in adenine nucleotides, tissue ATP, creatine phosphate, and improvement in cardiac function.

Preliminary electrophysiological studies with the tertiary MDIs demonstrated the ability of pr-MDI and bu-MDI to uncouple excitation-contraction coupling in superfused canine papillary muscle preparations at drug concentrations which do not reduce action potential characteristics including action potential amplitude, resting potential, duration at 25, 50, and 90% repolarization, and the rate of rise of phase 0 (59). This is taken as further evidence for an intracellular site of action of the MDIs. However, the electrophysiological effects of these agents appear to be quite complex (8,59).

Applied Pharmacological Studies with the MDIs

Calcium antagonists are known to decrease contractility of vascular smooth muscle and the myocardium (60). The value of this class of pharmacological agents in coronary therapeutics has been attributed to their ability to reduce myocardial oxygen consumption, decrease arterial blood pressure, and improve myocardial oxygen supply through dilation of extramural coronary arteries, collaterals, and anastamoses (61). The pr- and bu-MDIs ($5x10^{-6}M$ to $10^{-4}M$) produced a prolonged, concentration-dependent relaxation of potassium-depolarized strips of bovine extramural coronary vessels, which was reversible upon elevation of the calcium concentration of the medium (62). The effect was qualitatively similar to that observed with prenylamine (62) and other members of the prenylamine group (61) and dissimilar to the short-lived action of the nitrites (61). In the nonstimulated isolated perfused rabbit heart preparation, pr-MDI ($3x10^{-5}M$) and bu-MDI ($3x10^{-5}$ to $10^{-4}M$) increased coronary flow fourfold greater than the resulting decrease in inotropic activity, and did not alter chronotropic properties of the heart (62). Similar results are obtainable with the prenylamine group of drugs (61,62,63), many members of which enjoy extensive therapeutic use outside the USA in the management of angina pectoris of effort (63) as well as variant Prinzmetal angina (64).

The antiarrhythmic properties of the MDIs were investigated in several animal models. Ouabain-induced ventricular arrhythmias in dogs could be reversed by i.v. doses of 5-30 mg/kg of either pr-MDI or bu-MDI; the reversion being preceded by a drop in diastolic blood pressure, with no appreciable alteration in systolic pressure or heart rate (65). Pretreatment of dogs with pr- or bu-MDI also afforded significant protection against ouabain-induced arrhythmias (65) and calcium-induced arrhythmias (66). Pretreatment of rats with 3.75 mg/kg i.v. of either pr-MDI or bu-MDI provided virtually complete protection against calcium-induced bradycardia, arrhythmias, and death (66). The antiarrhythmic potency of the MDIs in rats was equivalent to that of verapamil, but the latter drug produced bradycardia, EKG alterations, and AV-block, whereas the MDIs exhibited no intrinsic deleterious effects on cardiac function (66). Prenylamine and phenytoin exhibited no protective action against calcium-induced arrhythmias in rats (66). Pretreatment of rats with the MDIs afforded significant protection against aconitine-induced arrhythmias, the order of potency being Q-bu-MDI (0.25 mg/kg, i.v.), quinidine (8 mg/kg, i.v.), bu-MDI (16 mg/kg, i.v.), and pr-MDI (24 mg/kg, i.v.) (67). Pretreatment of rats with the MDIs also afforded significant protection against methacholine-induced arrhythmias, the order of potency being Q-bu-MDI (1 mg/kg, i.v.), bu-MDI (2 mg/kg, i.v.), quinidine (4 mg/kg, i.v.), and pr-MDI (8 mg/kg, i.v.) (67). In the chloroform-anoxia assay, pretreatment of mice with the MDIs afforded at least a 70% protection against arrhythmias, the antiarrhythmic ED50s being: Q-bu-MDI (10.5 mg/kg, i.p.), bu-MDI (44 mg/kg, i.p.), quinidine (67 mg/kg, i.p.), and pr-MDI (68 mg/kg, i.p.) (67). The pronounced in-vivo antiarrhythmic activity observed with quaternary bu-MDI (Q-bu-MDI) in these studies may be attributable to possible in-vivo metabolic activation since this quaternary compound was relatively inactive in pharmacological in-vitro experiments performed so far (42,43). Alternatively, affinity for and accumulation in cardiac tissue may result in a sufficient concentration gradient to allow cellular penetration by diffusion or alteration of membrane permeability by the Q-bu-MDI, as has been proposed for the quaternary ammonium antiarrhythmic agents bretylium (68) and pranolium (69). The significant contribution of calcium antagonists to the therapy of cardiac arrhythmias has recently been reviewed (1,70,71).

In the heart, the action of histamine on H_1 receptors causes an increase in conduction time through the AV-node and an increase in coronary blood flow, whereas its action on H_2 receptors augments the force of contraction, increases cardiac rate and automaticity, and potentiates digitalis toxicity (72-76). H_1 and H_2 receptor antagonists have been shown to protect the heart against ouabain-induced arrhythmia (72,73) and to block the cardiac effects of histamine and other H_1 and H_2 agonists (74,75). Recently it has been demonstrated (77) that histamine-activated cardiac adenylate cyclase (which has the properties of an H_2 receptor) is blocked by the specific H_2 antagonist cimetidine ($pA_2 = 6.1$), as well as by the calcium antagonists L-cis-diltiazem ($pA_2 = 6.94$), verapamil ($pA_2 = 5.44$), perhexiline ($pA_2 = 5.58$), and pr-MDI ($pA_2 = 6.46$). The contribution of this H_2 blocking property of the calcium antagonists to their antiarrhythmic actions is not currently known since it neither corresponded to their calcium antagonistic potencies nor to their ability to block the positive inotropic action of histamine (77).

The effect of calcium antagonists on platelet function is very complex. Platelet activation by various stimuli appears to be initiated by the stimulus-induced translocation of intracellular calcium stores which sets in motion an integrated set of responses involving energy-producing reactions, contraction of actomyosin-containing filaments, alteration of microtubule assembly and distribution, release of granules containing stored products which facilitate aggregation and clotting, change in membrane properties leading to cellular adhesion, activation of the synthesis of PGs, PG endoperoxides, and thromboxane A_2, and the appearance of procoagulant platelet factor 3 in the surface membrane (78). The calcium antagonist TMB-8 has been shown to reduce platelet secretion and aggregation induced by thrombin, ionophores, and ADP (79,80). On the other hand, at concentrations up to 0.5 mM, pr-MDI induced platelet secretion and thus may act as a calcium agonist in this system (79). This latter action is reminiscent of the action of verapamil on skeletal muscle where the drug itself induces contracture and also potentiates caffeine-induced contractures (81). Nevertheless, pr-MDI does exhibit some calcium antagonistic actions on platelets in that it blocks thrombin-induced secretion (79).

Toxicological Studies with the MDIs

The acute and subacute toxicological profiles of the MDIs were recently reported from our laboratories (82). Acute toxicity studies resulted, in mice, in an i.v. LD50 of 40 and 32 mg/kg for pr-MDI and bu-MDI, respectively, and an i.p. LD50 of 185 mg/kg for both MDIs. In rats, the i.p. LD50 was 175 and 240 mg/kg for pr-MDI and by-MDI, respectively. In view of the antiarrhythmic equipotency of verapamil and the MDIs in the calcium-induced arrhythmia model in rats (66), it is clear that the therapeutic ratio of the MDIs is superior to that of verapamil since the latter exhibits an LD50 of 67 mg/kg (i.p.) and 16 mg/kg (i.v.) in rats and 68 mg/kg (i.p.) and 6.7 mg/kg (i.v.) in mice (83). Likewise, the therapeutic indices of the MDIs in the chloroform-anoxia assay in mice compare favorably with that of quinidine (67), the latter having an LD50 of 225 mg/kg (i.p.) (84). The LD50 of Q-bu-MDI in mice is 65 mg/kg (i.p.).

An i.v. dose of 16 mg/kg of pr-MDI or bu-MDI decreased motor activity and prolonged barbiturate sleeping time in mice, but did not affect conditioned avoidance behavior or motor coordination tests (82). In subacute toxicity studies (82), rats received daily for 4 weeks 26.25 or 52.5 mg/kg i.p. of each of the pr-MDI or bu-MDI, while mice received 23.13 or 46.25 mg/kg i.p. of either MDI. No alterations were observed in serum alkaline phosphatase, glutamic-pyruvic transaminase, glutamic-oxalacetic transaminase, creatine phosphokinase, bilirubin, chloride, cholesterol, uric acid, prothrombin time, and bromsulphalein retention. Blood glucose was slightly lowered, and so was serum calcium in male mice. The higher dose of pr-MDI elevated serum lactate dehydrogenase in rats. Both MDIs elevated serum isocitric dehydrogenase in male rats. However, light microscopic examination of brain, kidney, liver, spleen, intestine, stomach, and myocardium showed no anomalies resulting from the 4-week MDI treatment, and electron microscopic examination of hepatocytes revealed no deleterious effects of either MDI (82).

In an in-vitro study designed to investigate the effects of pr-MDI, bu-MDI, and Q-bu-MDI on the mechanical and electrical activity of the isolated guinea pig atria, particularly with regard to those effects which have positive or negative implications upon potential therapeutic usefulness of these agents as antiarrhythmic drugs, the following results were obtained (42): The tertiary pr-MDI and bu-MDI caused marked concentration-dependent decreases in contractile force of stimulated left atria, while the Q-bu-MDI caused only a slight peak reduction in contractile force. At a concentration of $3x10^{-4}M$ (which is higher than the maximum concentration of $10^{-4}M$ used in other pharmacological studies reported in this review), pr-MDI and bu-MDI significantly depressed the frequency-force profile of stimulated left atria, and significantly decreased membrane excitability (as reflected by a significant increase in threshold voltage), while Q-bu-MDI did not alter either parameter significantly. Spontaneous right atrial rate was only slightly depressed by the three MDIs at concentrations of $3x10^{-5}$ M or less. At higher concentrations, the tertiary pr- and by-MDIs were much more potent than the quaternary analogue in reducing atrial rate. These results indicate that Q-bu-MDI, the most potent in-vivo antiarrhythmic analogue of the MDI series, exhibits significantly lesser deleterious in-vitro effects on cardiac function than the tertiary MDIs (42). In turn, the tertiary MDIs exhibit far fewer deleterious electrocardiographic changes in-vivo than verapamil in equipotent doses (66). Nevertheless, it should be kept in mind that the in-vitro inactivity of Q-bu-MDI may be a reflection not of the greater safety of this analogue, but rather that the active in-vivo moeity may be a metabolite which is not formed in-vitro. The metabolic fate of Q-bu-MDI is one which requires investigation.

In isolated rabbit heart preparations, pr-MDI and bu-MDI ($3x10^{-5}$M) produced a negative inotropic action with no alteration in chronotropy (62). The negative inotropic action was sizably smaller than the increase in coronary flow produced by these agents (62). The decrease in force of contraction is beneficial for the purpose of conservation of oxygen in an ischemic heart. In-vivo studies in dogs (65) also demonstrated that the tertiary MDIs did not alter cardiac rate.

Structure-Activity Studies on the MDIs

Structure-activity studies done so far in our laboratories can be summarized as follows: shortening the 2-substituent (to produce 2-methyl- and 2-ethyl-3-dimethylamino-MDIs) resulted in reduction of calcium antagonistic activity (23). Lengthening the 2-substituent or rendering it more bulky (2-n-heptyl-, 2-cyclohexyl-, and 2-phenyl-3-dimethylamino-MDIs) resulted in reduction or loss of calcium-antagonistic activity and emergence of agonist activity (85). The agonist activity of the phenyl analogue on smooth muscle was not blocked by atropine (a muscarinic receptor blocker) or propyl-dibenzyloxyindanpropionic acid [a $PGF_{2\alpha}$ receptor blocker (21)], but was blocked by bu-MDI and by prenylamine, indicating an involvement of increased calcium influx (85). Opening the 5,6-methylenedioxy bridge (with resultant formation of the corresponding 5,6-dimethoxy analogues) results in reduction of calcium antagonistic properties and emergence of agonist activity (86).

The diastereoisomeric dihydro analogues of pr-MDI and bu-MDI (cis and trans 2-n-propyl- and 2-n-butyl-1-dimethylamino-5,6-methylenedioxyindanes) were synthesized and tested for their ability to reverse norepinephrine- or KCL-induced contraction of the isolated rat aorta (43). Saturation of pr-MDI and bu-MDI to produce the corresponding cis-aminoindane analogues yielded compounds with similar spasmolytic activity to the unsaturated parent compounds, while saturation which yielded the trans forms resulted in significant loss of potency (43).

N-Methylation of pr-MDI and bu-MDI and their cis and trans dihydro derivatives to produce the corresponding quaternary ammonium derivatives resulted in significant loss of spasmolytic activity against norepinephrine- and KCL-induced contractions of the rat aorta (43). Similarly Q-bu-MDI had little activity on the force of contraction of the isolated guinea pig atrium as compared to the tertiary compounds (42). This is to be anticipated due to the limited penetration of quaternary compounds into cells. Nevertheless, as mentioned earlier, Q-bu-MDI has potent antiarrhythmic activity in-vivo (67), possibly due to metabolic activation or accumulation in cardiac tissue.

Acknowledgment

Work from the authors' laboratories was supported by Grant HL-21670 from the National Heart, Lung, and Blood Institute of the U.S. National Institutes of Health.

Literature Cited

1. Rahwan, R. G., Piascik, M. F., Witiak, D. T. Canad. J. Physiol. Pharmacol., 1979, 57, 443-460.
2. Henry, P. D. Prac. Cardiol., 1979, 5, 145-156.
3. Elrodt, G., Chew, C.Y.C., Singh, B. N. Circulation, 1980, 62, 669-679.
4. Stone, P. H., Antman, E. M., Muller, J. E., Braunwald, E. Ann. Int. Med., 1980, 93, 886-904.
5. Zśoter, T. T. Amer. Heart J., 1980, 99, 805-810.
6. Sobel, B. E. Prac. Cardiol., 1981, 7, 31-46.
7. Rahwan, R. G., Witiak, D. T. "Trace Metals in Health and Disease", Kharasch, N., Ed., Raven Press, New York, 1979, p. 217.
8. Rahwan, R. G., Witiak, D. T., Muir, W. W. Ann. Rep. Med. Chem., 1981, 16, 257-268.
9. Rahwan, R. G., Med. Res. Rev., in press.
10. Sugiyama, S., Kitazawa, M., Kotaka, K., Miyazaki, Y., Ozawa, T. J. Cardiovasc. Pharmacol., 1981, 3, 801-806.
11. Crankshaw, D. J., Janis, R. A. Daniel, E. E., Canad. J. Physiol. Pharmacol., 1977, 55, 1028-1032.
12. Bleeker, A., Van Zweiten, P. A. Progr. Pharmacol., 1978, 2, 39-49.
13. Mikkelsen, E., Andersson, K. E., Pederson, O. L. Arch. Pharmacol. Toxicol., 1979, 44, 110-119.
14. Hidaka, H., Yamaki, T., Naka, M., Tanaka, T., Hayashi, H., Kobayashi, R. Molec. Pharmacol., 1980, 17, 66-72.
15. Kondo, K., Suzuki, H., Okuno, T., Saruta, T. Arch. Int. Pharmacodyn. Ther., 1980, 245, 211-217.

16. Church, J., Zśoter, T. T. Canad. J. Physiol. Pharmacol., 1980, 58, 254-264.
17. Akaike, N., Brown, A. M., Nishi, K., Tsuda, Y., Brit. J. Pharmacol., 1981, 74, 87-95.
18. VanBreemen, C., Hwang, O. K., Neisheri, K. D., J. Pharmacol. Exp. Ther., 1981, 218, 459-463.
19. Boström, S. L., Ljung, B., Mardh, S., Forsen, S., Thulin, E. Nature (London), 1981, 292, 777-778.
20. Witiak, D. T., Williams, D. R., Kakodkar, S. V., Hite, G., Shen, M. S., J. Org. Chem., 1978, 22, 77-81.
21. Witiak, D. T., Kakodkar, S. V., Johnson, T. P., Baldwin, J.R., Rahwan, R. G., J. Med. Chem., 1974, 39, 1242-1247.
22. Heaslip, R. J., Rahwan, R. G., Hassan, A.M.M., Witiak, D.T. Res. Comm. Chem. Pathol. Pharmacol., 1981, 32, 251-259.
23. Rahwan, R.G., Faust, M.M., Witiak, D.T., J. Pharmacol. Exp. Ther., 1977, 201, 126-137.
24. Piascik, M. F., Rahwan, R. G., Witiak, D. T., J. Pharmacol. Exp. Ther., 1978, 205, 155-163.
25. Rahwan, R. G., Borowitz, J. L., Miya, T. S., J. Pharmacol. Exp. Ther., 1973, 184, 106-118.
26. Rahwan, R. G., Borowitz, J. L., J. Pharm. Sci., 1973, 62, 1911-1923.
27. Rahwan, R. G., Borowitz, J. L., Hinsman, E. J., Arch. Int. Pharmacodyn. Ther., 1973, 206, 345-351.
28. Schneider, F. H., Biochem. Pharmacol., 1969, 18, 101-107.
29. Borowitz, J. L., Biochem. Pharmacol., 1969, 18, 715-723.
30. O'Neill, P. J., Rahwan, R. G., J. Pharmacol. Exp. Ther., 1975, 193, 513-522.
31. Rahwan, R. G., O'Neill, P. J., Miller, D. D., Life Sci., 1974, 14, 1927-1938.
32. Schneider, F. H., J. Pharmacol. Exp. Ther., 1971, 177, 109-118.
33. Rahwan, R. G., Gerald, M. C., Canad. J. Physiol. Pharmacol., 1981, 59, 617-620.
34. Bianchi, C. P., "Cellular Pharmacology of Excitable Tissue", Narahashi, T., Ed., Charles Thomas, Springfield, Ill., 1975, p. 485.
35. Burchfield, D. M., Rahwan, R. G., Rall, J. A., Amer. J. Physiol., in press.
36. Homsher, E., Mommaerts, W.F.H.M., Richiutti, N. V., Wallner, A. J. Physiol. (London), 1972, 220, 601-625.
37. Smith, I.C.H., J. Physiol. (London), 1972, 220, 583-599.
38. Loutzenhiser, R. D., Van Breemen, C., Fed. Proc., 1981, 40, 624.
39. Heaslip, R. J., Rahwan, R. G., Canad. J. Physiol. Pharmacol., in press.
40. Heaslip, R. J., Rahwan, R. G., J. Pharmacol. Exp. Ther., in press.
41. Heaslip, R. J., Rahwan, R. G., Fed. Proc., 1982, 41, 8397 Abst.

42. Lynch, J. J., Rahwan, R. G., Witiak, D. T. Pharmacology, in press.
43. Witiak, D. T., Brumbaugh, R. J., Heaslip, R. J., Rahwan, R.G. J. Med. Chem., in press.
44. Lynch, J. J., Rahwan, R. G., Canad. J. Physiol. Pharmacol., in press.
45. Bristow, M. R., Green, R. D., Europ. J. Pharmacol., 1977, 45, 267-279.
46. Zipes, D. P., Besch, H. R., Jr., Watanabe, A. M., Circulation, 1975, 51, 761-766.
47. Langer, G. A., J. Molec. Cell Cardiol., 1980, 12, 231-239.
48. Wollenberger, A., Will, H., Life Sci., 1978, 22, 1159-1178.
49. Beeler, G. W., Reuter, H., J. Physiol. (London), 1970, 207, 165-190.
50. New, W., Trautwein, W., Pflüger's Arch. Europ. J. Physiol., 1972, 334, 1-38.
51. Greenspan, A. M., Morad, M., J. Physiol. (London), 1975, 253, 357-384.
52. McDonald, T. F., Nawrath, H., Trautwein, W., Circ. Res., 1975, 37, 674-682.
53. Piascik, M. T., Johnson, C. L., Potter, J. D., Rahwan, R.G. Fed. Proc., 1981, 40, 1158 Abst.
54. Cheung, W. Y. Science, 1980, 207, 19-27.
55. Potter, J. D., Johnson, J. D., Dedman, J. R., Schreiber, W.E., Mandel, F., Jackson, R. L., Means, A. R. "Calcium Binding Proteins and Calcium Function", Wasserman, R.H., Corradino, R.A., Carafoli, E., Kretsinger, R.H., McLennan, D.H., Siegel, F.L., Eds., North Holland Publ. Co., New York, 1977, p. 239.
56. Klee, C. B., Crouch, T. H., Richman, P. G. Ann. Rev. Biochem., 1980, 49, 489-515.
57. Nagao, T., Matlib, M.A., Franklin, D., Vaghy, P. L., J. Mol. Cell. Cardiol., 1980, 12, 29-43.
58. Matlib, M. A., Vaghy, P. L., Johnson, J. D., Rebman, D., Rahwan, R. G., Schwartz, A., Proc. 8th Int. Congr. Pharmacol., 1981, Abst. 1025.
59. Muir, W. W., "Calcium Regulation and Drug Design", Rahwan, R. G., Witiak, D. T., Eds., Amer. Chem. Soc. Symp. Series, this volume.
60. Fleckenstein, A., "New Therapy of Ischemic Heart Disease", Lochner, W., Ed., Springer-Verlag, New York, 1975, p. 1.
61. Fleckenstein, A., Nakayama, K., Fleckenstein-Grün, G., Byon, Y. K., "Coronary Angiography and Angina Pectoris", Lichten, P. R., Ed., Publishing Sciences Group, Massachusetts, 1976, p. 297.
62. Piascik, M. T., Rahwan, R. G., Witiak, D. T., J. Pharmacol. Exp. Ther., 1979, 210, 141-146.
63. Dobbs, W., Povalski, H. J., "Cardiovascular Pharmacology", Antonacio, M. J., Ed., Raven Press, New York, 1977, p. 461.

64. Luchi, R. J., Chahine, R. A., Raizner, A. E. Ann. Int. Med. 1979, 91, 441-450.
65. Piascik, M. F., Piascik, M. T., Witiak, D. T., Rahwan, R.G. Canad. J. Physiol. Pharmacol., 1979, 57, 1350-1358.
66. Lynch, J. J., Rahwan, R. G., Witiak, D. T., J. Cardiovasc. Pharmacol., 1981, 3, 49-60.
67. Lynch, J. J., Rahwan, R. G. Fed. Proc., 1981, 40, Abst. 645.
68. Anderson, J., Patterson, E., Collin, M., Pitt, B., Lucchesi, B. R. Circulation, 1979, 60 (Suppl. 2), II-201 Abst.
69. Gibson, J. K., Korn, N. J., Counsell, R. E., Lucchesi, B.R. Fed. Proc., 1979, 38, 679.
70. Mason, D. T., DeMaria, A. N., Amsterdam, E. A., Vismara, L. A., Miller, R. R., Vera, Z., Lee, G., Zelis, R., Massumi, R. A. "Cardiovascular Drugs", Avery, G. S., Ed., ADIS Press, Sydney, 1977, p. 75.
71. Opie, L. H., Lancet, 1980, 2, 861.
72. Levi, R., Capurro, N. J. Pharmacol. Exp. Ther., 1975, 192, 113-119.
73. Somberg, J. C., Bounos, H., Cagin, N., Levitt, B. J. Pharmacol. Exp. Ther., 1980, 214, 375-380.
74. Sakai, K. J. Cardiovasc. Pharmacol., 1980, 2, 607-622.
75. Trzeciakowski, J. P., Levi, R. J. Pharmacol. Exp. Ther., 1980, 214, 629-634.
76. Laher, I., McNeill, J. H. Canad. J. Physiol. Pharmacol., 1980, 58, 1114-1116.
77. Johnson, C. L., Rupp, G., Pharmacologist, 1979, 21, Abst. 248.
78. Feinstein, M. B. "Calcium in Drug Action", Weiss, G. B., Ed., Plenum Press, New York, 1978, p. 197.
79. Murer, E. H., Stewart, G. J., Davenport, K., Siojo, E., Rahwan, R.G., Witiak, D. T. Biochem. Pharmacol., 1981, 30, 523-530.
80. Charo, I. F., Feinman, R. D., Detweiler, T. C., Biochem. Biophys. Res. Commun., 1976, 72, 1462-1467.
81. Bondi, A. Y. J. Pharmacol. Exp. Ther., 1978, 205, 49-57.
82. Rahwan, R. G., Akesson, C. E., Witiak, D. T. Res. Commun. Chem. Pathol. Pharmacol., 1979, 26, 85-103.
83. Haas, von H., Hartfelder, G. Arzneim. Forsch., 1962, 12, 549-558.
84. Lawson, J. W. J. Pharmacol. Exp. Ther., 1968, 160, 22-31.
85. Akesson, C. E., Rahwan, R. G., Brumbaugh, R. J., Witiak, D. T. Res. Commun. Chem. Pathol. Pharmacol., 1980, 27, 265-276.
86. Witiak, D. T., Kakodkar, S. V., Brunst, C. E., Baldwin, J. R., Rahwan, R. G. J. Med. Chem., 1978, 21, 1313-1315.

RECEIVED June 11, 1982.

Pharmacologic Consequences of Calcium Interactions with Opioid Alkaloids and Peptides

DAVID CHAPMAN [1] and E. LEONG WAY

University of California, School of Medicine, Department of Pharmacology, San Francisco, CA 94143

Considerable evidence implicates calcium ion in opiate action. The pharmacologic and neurochemical data include:

(1) Ca^{++} and its ionophores antagonize opiate action

(2) Ca^{++} antagonists (La^{+++} or EGTA) enhance opiate action

(3) Cross tolerance to La^{+++} and EGTA is exhibited by the morphine tolerant state

(4) The opiate abstinence syndrome can be alleviated by reducing neuronal Ca^{++}

(5) Acute opiate administration lowers neuronal Ca^{++}

(6) Chronic opiate administration elevates neuronal Ca^{++}

Based on our assessment of Ca^{++}-morphine interactions the role of Ca^{++} in acute and chronic opiate action is postulated to be as follows. There are two opposing effects of opiates on neuronal Ca^{++}, an immediate response to lower Ca^{++} and a delayed one which reflects counteradaptation to reverse the acute lowering effect on Ca^{++}. These two opposing actions of morphine can be utilized to explain its classic effects, on analgesia, tolerance and physical dependence. Our operational hypothesis is that the nociceptive state is regulated by the Ca^{++} level within the neuron, a lowering effects analgesia and an elevation in Ca^{++} hyperalgesia. The lowering of neuronal Ca^{++} induced by acute opiate administration is opposed by a homeostatic mechanism which tends to reverse the reduction in Ca^{++}. This latter process is cumulative so that with continuous opiate administration there is a

[1] Current address: University of Oxford, Department of Human Anatomy, Oxford 0X1 3QK, England

0097-6156/82/0201-0119$07.00/0

> gradual build up of neuronal Ca^{++}. The consequence would be tolerance development since to cause analgesia more opiate would be required to lower the elevated neuronal Ca^{++}. Under such conditions, a new elevated steady state for Ca^{++} becomes established whereby lowering of Ca^{++} becomes more difficult (tolerance) and the retention of the Ca^{++} requires the presence of opiate (physical dependence). With abrupt discontinuance of the opiate, there is loss of sequestered Ca^{++} and an increase in cytosol Ca^{++} ensues. The abstinence syndrome would then reflect a supersensitive state to opiate lack or a hyperirritable response to Ca^{++} excess that is ordinarily inhibited by the presence of opiate.

Numerous possible mechanisms of action at the biochemical level have been proposed, in order to explain the effects of opiates. In recent years the possibility that Ca^{2+} disposition is an important underlying site of opiate action has attracted widespread interest. This is particularly so in view of the fact that Ca^{2+} is involved in numerous aspects of neuronal function, a number of which are also known to be affected by opiates, and in particular, neurotransmitter release. The concept that Ca^{2+} represents a general site of opiate action has therefore arisen as a possible explanation for the varied behavioral and biochemical effects of these drugs.

Correspondingly, many papers have now appeared which concern metal ion (and especially Ca^{2+}) interactions with opiates, and each year a considerable number of new reports are published. This literature has been the subject of several reviews (1-4). In the present article this body of work is considered in terms of the general hypothesis that neuronal Ca^{2+} levels control the nociceptive state and that opioids, both endogenous and exogenous, exert their main effects via changes in Ca^{2+} disposition.

Ca^{2+} Antagonism of Opiate Effects

(a) *In vivo*. A number of reports have indicated that Ca^{2+} has inhibitory effects on *in vivo* opiate actions and especially analgesia. For example, Hano *et al*. (5) demonstrated a marked supression of morphine analgesia after intracisternal (i.c.) injection of Ca^{2+}. The Ca^{2+} and Mg^{2+} chelator ethylenediaminetetraacetic acid (EDTA) potentiated morphine analgesia, while EDTA alone produced a mild analgesia which was reduced by eqimolar doses of Ca^{2+}. Later the same laboratory (6) reported that i.c. Ca^{2+} administration antagonized the analgetic effects of morphine and it surrogates while other cations (Mg^{2+}, Ba^{2+}, Sr^{2+}, Zn^{2+}, Fe^{2+}, Ni^{2+}, K^{+} and Na^{+}) did not.

More recently it was reported by Harris *et al.* (7) that Ca^{++} antagonized morphine analgesia after intracerebroventricular injection, while Sr^{2+}, Ba^{2+}, Ni^{2+}, Hg^{2+} and Cd^{2+} were without effect. Although EDTA had slight or no significant effect on morphine analgesia, another chelator, ethylene glycol tetraacetic acid (EGTA), did cause significant potentiation of analgesia. EGTA is reported to have a much higher affinity for Ca^{2+} than Mg^{2+} (8) and thus these results were again interpreted as an involvement of Ca^{2+} rather than Mg^{2+} depletion in producing these effects. Furthermore, the ionophore X537A was found to significantly increase the antagonistic effect of a low dose of Ca^{2+}. X537A has been shown to increase membrane permeability to divalent cations (9), and thus the authors postulated Ca^{2+} antagonism of morphine at an intracellular site(s).

Consistent with the above findings, we have found that analgesia induced by β-endorphin, methionine-enkephalin and stress is also antagonized by Ca^{++}; EGTA again potentiated analgesia while the ionophore A23187 increased the antagonistic potency of Ca^{2+} (10).

The development of tolerance to the analgetic effects of opiates has generally been reported to be retarded by acute Ca^{2+} administration, as has dependence development (11-14) In other reports where tolerance development was not affected, it is likely that the dose of Ca^{2+} used were insufficient (7).

Harris and coworkers (15, 16, 17) have further examined the involvement of Ca^{2+} in opiate actions using the rare earth lanthanum. This trivalent ion competes very effectively with Ca^{2+} at cation binding sites due to its having a similar ionic radius but greater valency and has been reported to inhibit Ca^{2+} binding and movement across biological membranes (18). The results of these experiments are again consistent with an involvement of Ca^{2+} metabolism in analgesia, since similarities between the actions of La^{3+} and morphine were evident. Thus a subanalgesic dose of La^{3+} significantly potentiated morphine analgesia after intracerebroventricular (i.c.v.) injection, while at higher doses both La^{3+} itself and another Ca^{++} blocker, cerium produced analgesia. On a molar basis, La^{3+} was approximately one-tenth as potent as morphine. Surprisingly, the effects of La^{3+} were antagonized by naloxone but a flat dose-response curve was obtained.

The possibility that La^{3+} was producing its effects by inhibiting Na^{+} and K^{+} conductance, as it does in the lobster axon (19), is unlikely since i.c.v. or periaqueductal gray (PAG) injection of high concentrations of Ca^{2+} or local anesthetics (both of which inhibit Na^{+} and K^{+} fluxes) failed to produce analgesia (6,7,17).

A relationship further existed between morphine and La^{3+} in that animals tolerant to morphine exhibited cross tolerance to La^{3+}. However, the level of tolerance to La^{3+} was lower than for morphine. Also, La^{3+} was found to affect physical dependence as evidenced by its ability to suppress abrupt and naloxone-precipitated withdrawal jumping in mice treated chronically with a high dose of morphine. Finally, it was observed that the PAG region of the midbrain was the most sensitive site investigated for both morphine and La^{3+} analgesia. This is also the region of high opiate receptor density (20). Recently Reddy and Yaksh (21) have reported weak analgesia in rats after intrathecal administration of La^{3+}. In contrast to the earlier study (16). The La^{3+} effect was not naloxone reversible, although Ca^{2+} did reverse antinociception. Perhaps when La^{3+} does cause naloxone-reversible analgesia, it may be related to release of endogenous opiates although it should be noted that enkephalin release is Ca^{2+} dependent (22). Thus, any failure to effect naloxone antagonism of La^{3+} analgesia might thus be due to the diurnal fluctuations in endogenous opiate levels which have been reported (23). It is also possible that the different sites of drug administration may have given rise to the different effects of naloxone, with analgesia resulting from i.c.v. injection of La^{3+} involving opiate receptors in some way, but not that due to spinal action.

Caruso and Takemori (24) have reported Ca^{2+} antagonism of morphine-induced respiratory depression. The convulsive properties of morphine and opiate peptides have been attributed to the ability to reduce membrane Ca^{++} flux (25). On the other hand, Huidobro-Toro and Way (26) failed to show any antagonism by Ca^{2+} or Mn^{2+} of the hyperthermia induced by β-endorphin, following i.c.v. injection of both agents. It was postulated that since hyperthermia is a more sensitive response to β-endorphin than analgesia, possibly higher doses of cations would be necessary to produce significant antagonism. Unfortunately, it was not possible to perform the experiment, since higher doses of the cations alone produced marked hypothermia. Ziegelgänsberger and Bayerl (27) found that the *in vivo* activity of spinal neurons was depressed by phoretically applied morphine and Ca^{2+}. One can only conclude that while many actions of opiates involve Ca^{2+}, some may not.

(b) *In vitro*. Numerous reports indicate that Ca^{2+} can alter physiological responses to opiates in isolated tissue systems, for example, the ability of morphine to inhibit K^{+}-stimulated respiration in rat brain cortical slices (28,29). We subsequently found this to occur only in a low or Ca^{2+}-free medium (30).

Many studies using the guinea-pig ileum preparation have shown that Ca^{2+} is able to antagonize the inhibitory effects of opiates on electrically induced contractions (31-34, 36,37) Unlike the situation with analgesia, it appeared that Mg^{2+} and Mn^{2+} did not cause antagonism of opiate effects (31,33,36).

The inhibitory response to morphine in this preparation is established as being due to an inhibition of acetylcholine release which Ca^{++} can reverse (35). We have been shown that the inhibitory effect of morphine on acetylcholine release can be reduced by elevating the Ca^{++} concentration in the incubation bath (36).

The antimorphine action of Ca^{2+} has been reported to be both competitive and non-competitive. Our laboratory (34) Opmeer and Van Ree (32) and apparently Heimans (37) as well have reported a competitive-type antagonism. Later, we noted that with high concentrations of Ca^{++} a non-competitive antagonism occurred (36).

We have examined this apparent paradox further and have found that the type of antagonism observed is related to the length of time the tissue is in contact with the increased Ca^{2+} levels (10). Thus when the gut preparation was incubated in the presence of high Ca^{2+} concentrations for periods of one hour or more, a non-competitive antagonism was observed. Conversely, when Ca^{2+} levels were increased shortly before or after the addition of the opiate, a competitive antagonism was seen with parallel shifts of dose-response curves and maximal effects at high drug doses. These results would appear to correspond with the techniques used in the various reports mentioned above, in as far as the methods are given. It thus appears that the non-competitive effect is due to a storage of Ca^{2+}, presumably in neuronal elements, since opiates do not affect the muscle component in the guinea-pig ileum preparation (35). These results are consistent with the theory that opiates inhibits Ca^{2+} influx into cells. Thus when raised intracellular levels of Ca^{2+} are obtained, an opiate insensitive component is introduced, resulting in non-parallel shifts of dose response curves.

Recently Opmeer and Van Ree (33) have shown that the inhibition of the contractile response in the ileum which follows high-frequency stimulation and is presumably partially due to release of endogenous opiates (38) is also antagonized by increased Ca^{2+} concentrations. However, Opmeer and Van Ree also found that *in vitro* tolerance development was not affected when strips were incubated in Ca^{2+} free buffer containing EGTA. Thus it was proposed that although Ca^{2+} appears to be involved in the acute effects of opiates on the guinea pig ileum, it was much less important for *in vitro* tolerance development.

An alternative explanation for experiments where Ca^{2+} is used to overcome the effects of opiates, is that the ions, by promoting increased neurotransmitter release, are simply masking the opiate effect. The molecular events resulting from opiate receptor activation would thus still occur, but would be detected to a lesser degree. In this case, Ca^{2+} antagonism of opiates would thus not necessarily imply an opiate effect on Ca^{2+} metabolism. It is consistent with this interpretation that Ca^{2+} does not affect in vitro tolerance development in the guinea-pig ileum (33). However, as discussed previously (11-14) in most cases tolerance/dependence development in vivo is retarded by Ca^{2+} administration.

In an electrophysiological study using the mouse vas deferens, Bennett and Lavidis (39) examined the hypothesis that morphine acts by blocking the binding of Ca^{2+} to the presynaptic membrane X-receptor, thus inhibiting neurotransmitter release. When the excitatory junction potential due to a single nerve impulse was measured, it appeared from data analysis that morphine acts as a competitive inhibitor of the action of Ca^{2+} in promoting neurotransmitter release. The effects of morphine on synaptic potential during high- and low- frequency stimulation were investigated, and the results were again consistent with a Ca^{2+} antagonist action of morphine. Morphine acted in the same way as Mg^{2+}, a known competitive inhibitor of Ca^{2+} with a Ca^{2+} receptor complex (40).

Using the same preparation, Illes et al. (41) found that the inhibitory effects of normorphine were antagonized by increasing Ca^{2+} concentration in the buffer, by removing Mg^{2+} or by adding 4-aminopyridine. It was also observed that short trains of impulses (3 Hz) caused facilitation of response to stimulation, probably due to elevation of intracellular Ca^{2+} concentration. Consistent with the above results, it was found that the inhibitory effects of normorphine were inversely proportional the length of the train.

In other electrophysiological studies, however, Dingledine and Goldstein (42) and Williams and North (43) reported that opiates still produced inhibition of neuronal activity in Ca^{2+}-free buffer. It thus appeared that in these cases the opiates were not acting via altered transmembrane fluxes. Obviously, different preparations yield different effects to opioids and it is important to sort out and explain such discrepancies.

One of the most important pieces of evidence supporting a role for opioid action in Ca^{++} disposition is the finding that the slow Ca^{++} channel may be altered by an enkephalin analog. Mudge et al. (44) have reported that (D-ala^2) enkephalin amide (DAEA) inhibits Ca^{2+}-dependent, K^+-stimulated substance p release from sensory neurons grown in dispersed cell culture. Electrophysiological examination showed that DAEA and enkephalin decreased the duration and magnitude of the evoked Ca^{2+} action potential, suggesting that inhibition of substance P release may be a result of reduced Ca^{2+} entry via a direct effect on inward Ca^{2+} current.

Opiate Effects on Ca^{++} Content

The reports of Ca^{2+} involvement in opiate actions have lead to numerous examinations of narcotic effects on Ca^{2+}

disposition. In an early study on mice, Shikimi et al. (45) observed a significant decrease in whole brain Ca^{2+} content 30 minutes after a subcutaneous injection of 100 mg/kg morphine, an exceedingly large dose; 20 mg/kg had no effect. Also, the higher dose had no effect in morphine tolerant animals. Subsequently, Ross et al. (46) and Cardenas and Ross (47) reported significant, dose-dependent decreases in the Ca^{2+} content of each of 8 discrete brain regions following morphine treatment. These results are remarkable in that very large reductions in tissue Ca^{2+} content (up to 44%) were produced by pharmacological doses of morphine (up to 25 mg/kg) and the effects were nearly equal in all brain regions, although these extraordinary findings have not been reproduced by other investigators (48). However, subsequent studies of opiate effects on the enriched nerve ending fraction (synaptosomes) of brain homogenates revealed significant changes after acute and chronic morphine treatment.

When the Ca^{2+} content of sub-cellular fractions of rat cerebral cortices were examined following acute morphine treatment, it was found that significant depletions only occurred in the synaptosomal fractions (49,50). No changes in Na^{+}, K^{+} or Mg^{2+} content of any fraction were observed. Findings in the mouse largely substantiated the data obtained on the rat. Although Harris and co-workers found no alteration in Ca^{2+} content of mouse synaptosomes after 25 mg/kg of morphine, they were able to detect changes at intrasynaptosomal site(s) with a more sensitive technique in which $^{45}Ca^{++}$ was injected 6 hours before sacrifice so as to label brain Ca^{2+} stores Using this method it was found that significant decrease in mouse brain synaptic vesicle Ca^{2+} content resulted from morphine administration (51,52).

Further studies in our laboratory revealed that chronic treatment with morphine in both rats and mice, on the other hand, produced opposite effects to acute treatment, with significant increases in synaptosomal Ca^{2+} levels being observed (50, 51, 52). The increases were reported to be localized in the synaptic vesicle fraction (51,52) and synaptic plasma membrane fraction (SPM) (53). Naloxone treatment blocked the increase in Ca^{2+} levels, while naloxone-precipitated withdrawal resulted in a return to control Ca^{2+} levels within 15 minutes after injection (50). Both β-endorphin and methionine-enkephalin were also noted to cause Ca^{2+} depletion of synaptic vesicles and SPM after acute treatment (3).

We further postulated that if acute treatment resulted in decreased Ca^{2+} levels and chronic treatment produced increased levels, then sub-cellular preparations from such animals should have increased and decreased numbers of Ca^{2+} binding sites respectively. *In vitro* $^{45}Ca^{++}$ binding experiments substantiated this postulate, with acute morphine treatment resulting in an increase in high-affinity (10^{-7} -

10^{-5} M) binding sites on SPM, while chronic treatment caused a decrease in binding. Similar effects were seen for low-affinity binding sites (10^{-3} M) on synaptic vesicles (50). It was noted that the failure to observe changes in Ca^{2+} content of SPM fractions (51,52) was probably due to these being changes related to high-affinity sites with a low Ca^{2+} volume compared with the total Ca^{2+} measured. There was no change in binding to intact synaptosomes, suggesting that binding sites on the inner surface of synaptic membranes were affected. In similar experiments decreased $^{45}Ca^{2+}$ binding capacity in SPM fractions were confirmed (53) but both high- and low-affinity sites were affected.

Taken together, the observed changes in synaptosomal Ca^{2+} content suggest that opiates may act by altering Ca^{2+} binding and/or fluxes at these sites. This would then in turn lead to changes in the activity of Ca^{2+}-dependent enzymes and ultimately to altered patterns of neurotransmitter release.

In addition to the above experiments showing that *in vivo* opiate treatment causes changes in subsequent *in vitro* $^{45}Ca^{++}$ binding, there is also evidence that opiates may affect $^{45}Ca^{2+}$ binding by *in vitro* treatment. A dose-dependent inhibition by levorphanol of high-affinity $^{45}Ca^{2+}$ binding *in vitro* to synaptic membranes was reported; the inhibition was stated to be non-competitive, stereospecific and naloxone reversible (54). Changes in the shape of the Ca^{2+} binding curve from sigmoid to hyperbolic in the presence of the opiate, lead the authors to speculate that Ca^{2+} and opiate receptor sites are close together and may be linked through sub-unit interactions. In addition, levorphanol displaced previously bound $^{45}Ca^{2+}$ from membrane preparations. In a review article, the same authors claimed similar effects with β-endorphin (3). However, two reports concerning opiate effects on Ca^{2+} binding *in vitro* are not in agreement. Kaku *et al.* (55) observed no effect by morphine on $^{45}Ca^{2+}$ uptake (presumably binding) by mouse synaptic membrane fractions and Hoss et al.(56) reported morphine to be without effect on Ca^{2+} binding to synaptic vesicles (as determined from protein-sensitized fluorescence of Tb^{3+}) after *in vitro* or acute treatment.

Opiate effects on Ca^{2+} fluxes have been shown in a number of different experimental preparations. Kakunaga (57) reported that a high concentration of morphine (10^{-3} M) inhibited K^{+} and EDTA stimulated $^{45}Ca^{2+}$ influx and efflux in rat brain slices. The effect was partially nalorphine reversible, was seen in low-Ca^{2+} medium (0.1 mM), but not at a higher Ca^{2+} concentration (1.3 mM). We have re-examined these phenomena and have found naloxone-reversible morphine inhibition of K^{+}-stimulated $^{45}Ca^{2+}$ uptake into brain slices from cortex or midbrain, but not cerebellum (10). Unlike the earlier report this effect was observed at Ca^{2+} concentrations up to 1.4 mM. These results were not obtained below morphine concentrations of 5 x 10^{-4} M, suggesting opiate action via a low-affinity receptor or loss of co-factors requisite for the process.

Kaku et al. (55) reported that acute *in vivo* or *in vitro* morphine treatment inhibited ^{45}Ca uptake into mouse brain synaptosomes, an effect which disappeared with tolerance development. A series of experiments have been conducted in this laboratory in order to examine opiate effects on ^{45}Ca fluxes in synaptosomes. Morphine caused dose-dependent decreases in synaptosomal ^{45}Ca uptake at low K^+ concentrations (in the presence of ATP and Mg^{2+}) after *in vitro* or acute *in vivo* treatment (58). This effect was naloxone reversible and stereospecific, with levorphanol also causing inhibition but not dextrorphan. Uptake was apparently inhibited in a non-competitive fashion. In contrast, mice made tolerant by morphine pellet implantation showed progressive increases in ^{45}Ca uptake with increasing degrees of tolerance that had developed. β-endorphin similarly inhibited ^{45}Ca uptake into synaptosomes after *in vitro* treatment or after i.c.v. injection (59).

Subsequently it was shown that if a synaptosomal preparation was lysed, pelleted and resuspended, the uptake of ^{45}Ca by the intrasynaptosomal particles was again inhibited by prior acute *in vivo* or *in vitro* treatment (60). Similarly, an increased uptake again followed tolerance development. These drug effects were seen when ATP (3 mM) was included in the media but not in its absence, suggesting that an active process was being affected. It is unlikely that mitochondrial ^{45}Ca fluxes were altered by morphine, since the effects were observed in the presence of mitochondrial inhibitors. It was considered most likely that the site of action was the synaptic vesicles.

More recently, after allowing synaptosomes to reach a steady state ^{45}Ca uptake in high Na^+ medium (61), naloxone-reversible inhibition of basal ^{45}Ca uptake by morphine and β-endorphin could be consistently demonstrated (62,63) This inhibitory effect was noted only when ATP was present in millimolar concentrations in the incubation medium. Efflux of ^{45}Ca from preloaded synaptosomes was not significantly affected by morphine. The, inhibitory effect on Ca^{++} uptake was seen with synaptosomes prepared from frontal cortex, thalamus and hypothalamus, but not from the cerebellum, thus corresponding well with data on opiate receptor distribution (20,64).

Similar effects were stated to occur with rabbit synaptosomes, levorphanol causing a reduction of K^+-stimulated ^{45}Ca uptake *in vitro* (65). Once again this was a stereospecific, naloxone-reversible and non-competitive antagonism. Also, chronic levorphanol treatment of mice produce increases in K^+-stimulated uptake compared with controls. In this case no effects were seen with resting levels of ^{45}Ca flux at 5 mM K^+ (3). Very similar results were reported by End *et al.*

(66). It was not mentioned in these reports if ATP was included in the media. An absence of ATP may account for the failure to see effects at low K^+ concentrations. In other experiments End *et al.* (67) have demonstrated morphine effects on ^{45}Ca efflux from mouse neuroblastoma cells.

Calcium Hypothesis of Opiate Action

It is evident that a wide range of experimental data indicates that Ca^{2+} generally inhibits acute effects of opiates *in vivo* and also *in vitro*. In addition opiates produce a rapid fall in neuronal $Ca2^+$ levels after acute treatment, with a corresponding gradual increase in Ca^{2+} levels after prolonged treatment. These phenomena have been used to construct an hypothesis which provides an explanation for the classic effects of opiates, namely acute effects, such as analgesia, followed by tolerance and physical dependence development (68, 69). In this hypothesis it is assumed that the nociceptive state of an organism is controlled by neuronal Ca^{2+} levels.

Acute opiate treatment is thus considered to cause decreased Ca^{2+} binding and/or fluxes at nerve endings resulting in reduced neurotransmitter release, since release is dependent upon Ca^{2+} influx (70,71). As shown in Fig. 1 this could give rise to analgesia and other acute drug effects and also cause the decreased Ca^{2+} content observed in synaptic vesicles and SPM. In support of this we have already provided data that Ca^{2+} antagonists such as La^{3+} and EGTA can themselves produce analgesia and also potentiate opiate analgesia at low doses (15, 16, 17, 72). These agents are less potent analgesics than opiates, possibly because they are less selective and act on a wide variety of sites. Opiates appear to act at specific Ca^{2+} pools associated with SPM and synaptic vesicles. Conversely, Ca^{2+} not only can antagonize opiate-induced analgesia but can also produce hyperalgesia. Thus, the ionophores X537A and A23187, which probably act by facilitating Ca^{2+} influx, enhance Ca^{2+} antagonism of opiate action.

With prolonged narcotic treatment a homeostatic mechanism increasingly take over to retain Ca^{2+}, which results from increased Ca^{2+} binding and/or uptake at synaptic vesicle and SPM sites (Fig. 2). The elevated Ca^{2+} necessitates more narcotic to produce acute opiate effects, i.e. more is required to again reduce intracellular Ca^{2+} and produce analgesia. Therefore, tolerance develops as higher doses of opiate are required to produce an effect, and this in turn leads to further adaptation and an increasing cellular Ca^{2+} accumulation. One might anticipate that the elevated Ca^{++} would reduce the analgetic effects of La^{3+} and EGTA, and the cross-tolerance to these two agents noted in the morphine tolerant state (15, 16) is consistent with such expectations.

The augmented Ca^{++} cumulation requires the continual presence of the opiate, and so physical dependence development also occurs as Ca^{2+} levels rise. Thus when opiate discontinuance or antagonist treatment removes the agonist, the high synaptosomal Ca^{2+} content in the absence of the opiate produces greatly increased neurotransmitter release (Fig. 3) This neuronal hyperexcitability then gives rise to the classic withdrawal signs and symptoms. Therefore the raised Ca^{2+} levels observed during tolerance-dependence development should result in hyperalgesia after morphine at receptor sites has disappeared and this was verified experimentally. With decreased sensitivity to opiate there could be increased responsivity to Ca^{++} and this was noted to be the case (72). In fact, tolerant mice were significantly more sensitive to Ca^{2+} induced hyperalgesia than placebo controls. According to this hypothesis, reducing intracellular Ca^{2+} should attenuate the abstinence syndrome. In accordance with these expectations, La^{3+} administration reduced abrupt or naloxone-induced withdrawal jumping in mice (16).

This model would suggest that Ca^{2+} administration, by inhibiting opiate effects, should reduce tolerance development and such results have been noted previously (11-14). Furthermore, EGTA (due to its Ca^{2+} depleting effects) should enhance tolerance development, and again, such an effect has been reported by Schmidt (73).

Biochemical Interactions

Lipid metabolism. Although a model has been provided to explain the acute and chronic actions resulting from opiate treatment, it is not currently possible to identify precisely the molecular site(s) of opiate-Ca^{2+} interactions. Clearly the most obvious sites of action are those where Ca^{2+} metabolism has been reported to be modified by opiates e.g. Ca^{2+} binding. Also, there is as yet no definite evidence as to which mechanisms modulate the switch from acute effects to tolerance development.

One possible form of Ca^{2+}-opiate interaction is drug-induced inhibition of Ca^{2+} binding at synaptic membrane sites. However, as we have discussed previously the evidence for such an effect *in vitro* is equivocal. Mulé (74) has proposed that opiates displace Ca^{2+} from anionic binding sites on phospholipid molecules in neuronal membranes. The displacement of Ca^{2+} from these phospholipid opiate receptors would thus result in changes in membrane permeability to other ions, producing alterations in neuronal excitability.

Mulé (75) has reported narcotic inhibition of phospholipid-facilitated transport of $^{45}Ca^{2+}$ between Chloroform and water phases. However, this effect was shared by naloxone and dextrorphan, as well as non-narcotic drugs and potency was probably related to the degree of drug ionization. Greenberg

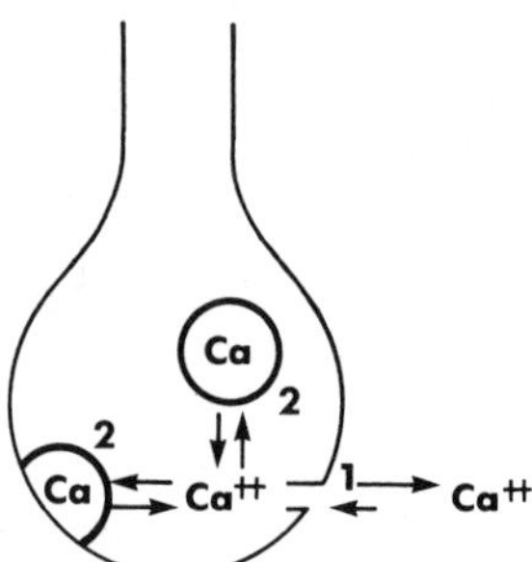

Figure 1. Postulated role of Ca^{2+} in acute opiate action.

Effects: Ca^{2+} uptake inhibited at 1, Ca^{2+} binding reduced at 2, synaptosomal Ca^{2+} decreased, cytosol Ca^{2+} decreased, no neurotransmitter release, counteradaptive response at 2 is initiated to retain Ca^{2+}.

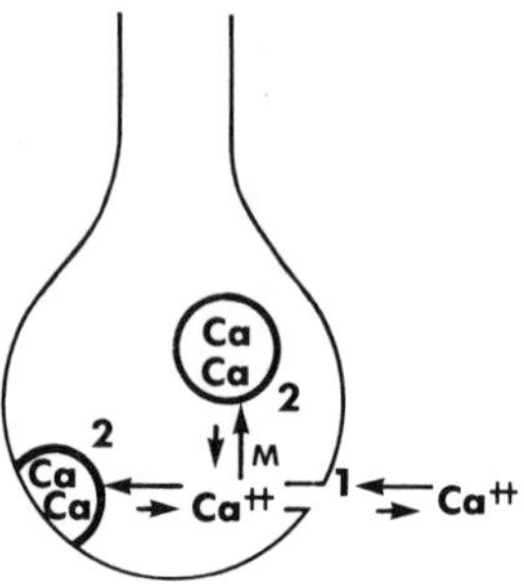

Figure 2. Postulated role of Ca^{2+} in chronic opiate action.

Effects: Tolerant-dependent state, Ca^{2+} uptake increased at 1 and 2, synaptosomal and cytosol Ca^{2+} elevated and opiate dependent, high dose of opiate required to block uptake at 1 and 2 (tolerance) and consequence is increased ability to cumulate Ca^{2+} at 2, response to opiates at 2 now more than at 1 but opiate dependent.

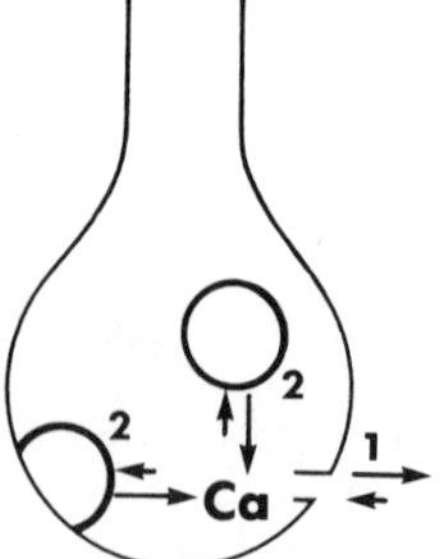

Figure 3. Postulated role of Ca^{2+} in withdrawal behavior.

Effects: Abstinence, inability to retain Ca^{2+} at 2, synaptosomal Ca^{2+} decreased, cytosol Ca^{2+} increased—irritability increased Ca^{2+} loss at 2 readily reversible with opiates.

et al. (76) later showed that morphine inhibited in vitro binding of $^{45}Ca^{++}$ to purified bovine gangliosides (acidic glycolipids occurring in neuronal membranes). The antagonist nalorphine had a biphasic effect, causing increased binding at lower doses and decreased binding at higher doses. The effects of nalorphine on morphine inhibition appeared to be complex, with no clear antagonism.

Based on data that the phospholipid base-exchange reaction in nervous tissue is stimulated by Ca^{2+} and is energy dependent (77); the more recent findings by Natsuki et al. (78) may have greater relevance. These authors postulated that morphine may alter the turnover and/or composition of membrane phospholipids by a direct effect on the Ca^{2+}-dependent base-exchange reaction. Acute and chronic morphine treatment were found to increase basal exchange (no Ca^{2+} added) of (^{14}C) serine in brain microsomal membranes from rats, while chronic but not acute treatment increased Ca^{2+}-stimulated exchange. Morphine in vitro also caused increases in basal and Ca^{2+}-stimulated exchange that could be reversed by naloxone. There was also some evidence of increased (^{14}C) ethanolamine exchange after chronic treatment, while (^{14}C) choline exchange was decreased at all Ca^{2+} concentrations tested. It is quite possible that the marked changes in exchange observed after chronic treatment represent a homeostatic mechanism, to overcome acute changes in membrane properties. Thus an increased serine turnover would also result in increased Ca^{2+} turnover, thereby antagonizing acute inhibitory effects on Ca^{2+} binding and/or uptake.

Evidence that membrane phospholipids are involved in Ca^{2+} gating mechanisms, such as the recent report by Putney et al. (79), are consistent with the notion that opiates acutely alter Ca^{2+} fluxes by inhibiting Ca^{2+} binding to phospholipids. However, the known actions of opiates at specific brain sites needs to be reconciled with the widespread occurence of membrane phospholipids. Recently work by Loh and co-workers (80,81) has shown that the glycolipid, cerebroside sulfate (CS) shows many of the properties of an opiate receptor. Although the ubiqitous distribution of the glycolipid appears to be inconsistent with the specific pattern of opiate receptor distribution, it is argued that only the cerebroside sulfate stragetically localized at membrane sites is critical. Hence cerebroside sulfate (or other glycopids) in association with certain unknown proteins may well represent a functional opiate receptor, which not only would give specificity to opiate-receptor binding but also to opiate-Ca^{2+}-lipid (cs) interactions.

There is considerable evidence to support the supposition that cerebroside sulfate might participate in opiate action. The data have been derived not only from in vitro experiments but in vivo ones as well and these papers have been cited and summarized (80). In brief, cerebroside sulfate fulfills the structural requirements of the theoretical models postulated for the opiate receptor. The putative opiate receptor isolated from homogenates of mice brain was identified to be cerebroside sulfate. Binding of opiates with cerebroside sulfate has been demonstrated to be of high affinity and stereospecific. Moreover, the concentration of various opiates required to inhibit radioactive opiate binding to cerebroside sulfate correlates well with their analgetic activity. Under conditions where the availability of cerebroside sulfate is decreased, there is reduced responsivity to opiates. Thus, the genetic mutant jimpy mice, which are known to have low brain content of cerebroside sulfate, were found to be less sensitive to morphine with respect to analgetic activity and to have lower binding capacity for morphine than their litter mate controls. Moreover, i.c.v. injection of the cationic dye, Azure A to selectively bind cerebroside sulfate in the brain also decreases morphine binding and analgetic potency. While most of the above evidence is circumstantial and indirect, taken together it argues strongly for a role for cerebroside sulfate (or other glycolipids) in opiate action.

Based on these and lipid metabolism data, Loh (81) has proposed a model to account for both acute and chronic opiate effects. It is suggested that opiate receptor binding results in acute effects due to change in membrane lipid structure. Since phospholipids are known to cause *in vitro* stimulation of gene expression it is further argued that prolonged drug treatment and the consequent changes in lipid metabolism, may give rise to the altered protein synthesis necessary for tolerance and physical dependence development. It has been shown that both syndromes, can be simultaneously inhibited by protein-synthesis inhibitors (82). This model is clearly in accord with the Ca^{2+} model described earlier since the changes outlined here could well involve changes in Ca^{2+} metabolism for both acute and chronic treatment.

ATPase

ATPase enzymes represent another possible site of opiate interference with Ca^{2+} flux, since these enzymes have an important function in active ion transport (83). A number of investigators have reported positive effects of opiates on ATPase activity after *in vitro*, acute and chronic treatment but no clear pattern emerges from these studies that can adequately explain either acute effects or tolerance development (55, 84-89).

Studies of opiate effects on ATPase activities have been conducted both by us and others, as a direct consequence of the earlier results showing opiate-induced alterations in Ca^{2+} levels. Synaptic vesicles are known to contain neurotransmitters, while Ca^{2+} is important for neurotransmitter release (71). Thus the changes in synaptic vesicle Ca^{2+} content after opiate treatment (51, 52, 53) may represent an interference with release mechanisms.

Following along these lines Yamamoto et al. (89) examined the effects of chronic morphine treatment on the Mg^{2+}-dependent ATPase of synaptic vesicles. Since this enzyme has been implicated in the regulation of neurotransmitter release (90,91), it appears a likely enzyme candidate to be altered by opiates. Consistent with this anticipation, it was found that the activity of Mg dependent ATPase in mouse brain synaptic vesicles was significantly increased with tolerance development while that of the Mg^{2+} dependent ATPase and Na^+, K^+ activated ATPase from SPM fractions were not altered. Kendrick et al. (92) have shown that Mg^{2+}-ATPase may be involved in the accumulation of Ca^{2+} by synaptic vesicles; the increased Ca^{2+} content of vesicles seen in tolerant animals could thus arise from increased enzymatic activity. Such a homeostatic mechanism would serve to overcome reduced levels of neurotransmitter release due to the opiate.

Other possible opiate activation sites could involve certain ATPases associated with calmodulin. A growing literature suggests that such Ca^{2+}-dependent regulator proteins (calcium dependent regulator) regulate the activity of a number of enzymes such as phosphodiesterase (93) and adenylate cyclase (94) via the formation of Ca^{2+}-CDR -enzyme complexes in response to Ca^{2+} fluxes. Thus, they appear to represent a link between different types of cell messenger, namely Ca^{2+} and cAMP. It has further been postulated that calmodulin, a CDR protein, is a likely Ca^{2+} receptor site (95). These proteins may thus represent an important site for Ca^{2+}-opiate interactions, with consequent alteration of enzyme activity.

The existing data on the effect of opiates on calmodulin and Ca^{2+} ATPase are confusing. A CDR protein isolated from heat treated synaptosomal lysates, presumably calmodulin, was found to stimulate Ca^{2+} ATPase activity. Morphine also stimulated Ca^{2+} ATPase but inhibited Ca^{2+} binding to this protein (96). It is also reported from the same laboratory that opiates inhibit Ca^{2+} ATPase activity in synaptic membranes (3). On the other hand, Levin and Weiss (97) report no effect by morphine on Ca^{2+} specific binding to calmodulin. Clearly, more work is needed to clarify these discrepancies.

Adenylate Cyclase and Guanylate Cyclase

Because both adenylate cyclase and guanylate cyclase appear to be Ca^{2+}-dependent enzymes acting probably via CDR proteins (98,99), several investigators examined the possibility that opiate actions on these enzymes are related to changes in Ca^{2+} metabolism. Thus it has been shown that Ca^{2+} must be present in the medium before opiates can induce changes in adenylate or guanylate cyclase activity (100-102). Adenylate cyclase activity in neuroblastoma x glioma cell lines was considered to be altered in a manner analagous to *in vivo* effects resulting in tolerance and dependence. Thus, while short-term administration results in enzyme inhibition, prolonged treatment induces increased activity once the restraint imposed by the drug has been removed (103). This system has been proposed as a model for studying opiate effects related to tolerance/dependence production, and has been employed recently in order to investigate the Ca^{2+}-dependence of opiate actions.

In a brief report, Wilkening (104) reported that morphine inhibited both basal and Ca^{2+}-induced adenylate cyclase activity in hybrid cells. Brandt *et al.* (105) have shown that low Ca^{2+} levels and opiates produce similar effects on adenylate cyclase response to prostaglandin E_1 (PGE_1) in neuroblastoma x glioma cell lines. Thus short-term (minutes) treatment produced decreased responses and long-term (hours) treatment produced enhanced responses. However, it appeared that opiate and low Ca^{2+} gave similar effects due to their similar action on cAMP levels rather than because of direct Ca^{2+}-opiate interaction in altering enzyme activity. Thus leucine enkephalin reduced PGE_1 stimulated cAMP levels even in the absence of Ca^{2+} from the external medium Furthermore, opiate activity was not altered when the ionophore A23187 was used to increase intracellular Ca^{2+} levels. However, the authors urged caution in considering these results as being at variance with those that support involvement of Ca^{2+} in opiate actions, pointing out that the hybrid cells differ considerably from normal neuronal cells, for example in terms of opiate receptor types.

Protein Kinase

Synaptic membrane-bound protein kinase activity is regulated by both cAMP and Ca^{2+} (106,107). Opiates might thus alter protein kinase activity by reducing Ca^{2+} availability, either directly or indirectly via adenylate cyclase. Enzyme induced protein phosphorylation is reported to cause altered membrane ion permeability and thus changes in neuronal excitability. Direct opiate effects on the enzyme could thus produce changes in Ca^{2+} distribution such as those discussed previously.

Clouet *et al.* (108) reported that both methadone and morphine inhibited A23187-stimulated phosphorylation of SPM in rat striatal synaptosomes; later the same workers showed that *in vitro* phosphorylation was decreased in SPM from morphine tolerant rats (109). Morphine *in vitro* did not affect protein kinase activity and although high doses of morphine did inhibit Ca^{2+}-stimulated phosphorylation, the effect was not naloxone reversible. More recently, Clouet and Williams (110) have shown that acute treatment of rats with a number of narcotics and opiate peptides results in an initial increase, followed by a subsequent decrease in striatal SPM phosphorylation *in vitro*. This pattern resembled that due to Ca^{2+} changes and it was postulated that the opiates, by reducing intrasynaptosomal Ca^{2+} levels, first gave rise to optimal Ca^{2+} levels (and hence increased phosphorylation), while further reduction of Ca^{2+} levels resulted in reduced phosphorylation. In preliminary communications, it was reported that Ca^{2+}-dependent (but not Ca^{2+}-independent) phosphorylation of synaptic protein was reduced by acute morphine treatment (111) and that Ca^{2+} and A23187-stimulated phosphorylation of synaptic proteins *in vitro* was enhanced in preparations from morphine tolerant mice (112).

Neurotransmitter release

A number of investigators have attempted to demonstrate that the well-estalished ability of opiates to inhibit neurotransmitter release is related to opiate-Ca^{2+} interactions.

Shikimi *et al*. (113) observed that Ca^{2+} was necessary in the incubation medium before high concentrations of morphine (10^{-3} M) could inhibit K^+-stimulated acetylcholine release from cerebral cortical slices. Also morphine inhibition of acetylcholine release from the cerebral cortex *in vivo* is antagonized by both subcutaneous (114,115) and intraventricular injection of Ca^{2+} (116) as well as in the guinea pig ileum (36).

Göthert *et al* (117) have reported that Ca^{2+} (1.3 mM) evoked overflow of norepinephrine in the presence of high K^+, was inhibited by methionine enkephalin. However, Ca^{2+}-promoted overflow in the presence of A23187 was not inhibited by the opiate. We have observed a similar failure of morphine to inhibit A23187-stimulated release of norepinephrine (10). The ability of the ionophore to transport Ca^{2+} across membranes is probably due to lipid soluble ionophore/Ca^{2+} complex formation (118). Thus the failure to affect ionophore-dependent release, strongly suggests that opiates act by inhibiting Ca^{2+} flow across membranes via potential-sensitive ion channels. Göthert and Wehking (119) have also shown that morphine inhibitory effects on norepinephrine release from rat occipital cortex slices are overcome with increasing Ca^{2+} concentration in the medium. However, this may not indicate a direct Ca^{2+}morphine interaction, since the large increases in neurotransmitter release resulting from the increased Ca^{2+} levels may simply have been masking the opiate effects. We have observed that morphine effects on *in vitro* norepinephrine release are only reduced while Ca^{2+} increases result in increased neurotransmitter release, but not when a maximum release level is attained (10). This would tend to argue against a direct Ca^{2+}-opiate involvement although a non-competitive inhibition of Ca^{2+} uptake is compatible with these results.

Down and Szerb (120) and Szerb (121) have examined the kinetics of evoked release of ^3H-acetylcholine from the guinea pig longitudinal muscle-myenteric plexus preparation both in the presence and absence of morphine. Stimulation produced an initial fast phase of release, followed by a slower phase. Morphine reduced both the size of the pool and rate of release of the fast efflux, but only the size of the pool of the slower efflux. It was suggested that the effect of morphine on slow release was due to hyperpolarization and not to any change in excitation-release coupling such as decreased Ca^{2+} influx, since this latter would mainly reduce the rate of release. Furthermore, low Ca^{2+} and oxotremorine depress neurotransmitter release due to reduced Ca^{2+} influx into terminals (122,123) and both were observed to reduce the rate of release but not the pool size. In contrast to these results Sawynok and Jhamandas (124) reported that morphine inhibition of acetylcholine release from the longitudinal muscle-myenteric plexus preparation was inhibited by an increased Ca^{2+} concentration which itself did not increase acetylcholine release. This perhaps suggests a direct opiate-Ca^{2+} interaction. Szerb (121) offers an alternative explanation however, suggesting that the the increased Ca^{2+} both hyperpolarizes neurons so that morphine can not hyperpolarize them further, and also accentuates acetylcholine release per action potential due to increased Ca^{2+} entry, resulting in little overall change in release.

Conclusion

The cited literature indicates that there is much evidence in favor of the suggestion that opiate drug effects involve Ca^{2+} disposition in some way. However, the multiple, and mutually interdependent actions of Ca^{2+} in neuronal function make it difficult to establish at which sub-cellular site or sites significant opiate-Ca^{2+} interactions are occurring. This is especially true since a number of Ca^{2+}-dependent effects are known to be susceptible to opiate action. The picture is further complicated by the possibility that in some cases opiate effects may not involve Ca^{2+}, while in the instances where opiate-Ca^{2+} interactions can be demonstrated, the effects on Ca^{2+} may be either directly related to drug action or may be the indirect consequence of other opiate actions such as effects on body temperature, pH or oxygen tension.

If altered Ca^{2+} metabolism is the central and significant feature of opiate agonist actions, this could be brought about in one of two ways. Firstly, the drug could act at neuronal sites to produce altered activity of enzymes (adenylate cyclase, protein kinase, ATPase), or lipid turnover and thereby lead to changes in Ca^{2+} disposition (e.g.Ca^{2+} flux) and drug effects. Secondly, the drug might directly interact with Ca^{2+} to produce its effects e.g. by inhibiting Ca^{2+} binding.

Alternatively, it may be argued that the opiate induced changes in Ca^{2+} flux, may be secondary to altered neuronal activity caused by other more important opiate actions which do not directly involve Ca^{2+} metabolism. Similarly, it may be that while Ca^{2+} has often been observed to antagonize opiate effects, this only indicates that the Ca^{2+} can alter levels of neuronal activity but does not prove a direct effect of opiates on Ca^{2+}. Since these two differing interpretations may be placed on much of the data, it would seem that an important objective for future work in this field is to instigate research which will effectively demonstrate whether or not opiate effects are being produced as a direct consequence of drug effects on Ca^{2+}.

If opiates and opiate peptides are acting on Ca^{2+} disposition they are not unique in this respect. Neurotransmitter release is known to be a Ca^{2+}-dependent process, and, in fact a number of neurotransmitters or pharmacologic agents tors such as norepinephrine, adenosine and oxotremorine have been reported to act by inhibition of Ca^{2+} influx (123,125,126). This must mean that opiates produce their characteristic spectrum of effects, not because they have a unique mode of neuromodulatory action, but because the receptors for the endogenous opiate ligands are located stratgetically at selective neuronal sites for the mediation of their specific effects.

Acknowledgements

The studies cited from our laboratory were largely supported by a research grant (DA 00037) from the National Institute on Drug Abuse.

Literature Cited

1. Kaneto, H. "Narcotic drugs: Biochemical pharmacology", Ed. D.H. Clouet, Plenum Press, New York, 1971, pp. 300-309.
2. Sanghvi, I.S. and Gershon, S. Biochem. Pharmacol. 1977, 26, 1183-1185.
3. Ross, D.H. and Cardenas, H.L. "Neurochemical mechanisms of opiates and endorphins", (Adv. Biochem. Pharmacol. Vol. 20), ed. H.H. Loh and D.H. Ross, 1979, pp. 301-306.
4. Chapman, D.B. and Way, E.L. Ann. Rev. Pharmacol. Toxicol., 1980, 20, 553-579.

5. Hano, K., Kaneto, H. and Kakunaga, T. Jap. J. Pharmacol. 1964, 14, 227-229.
6. Kakunaga, T., Kaneto, H. and Hano, K. J. Pharmacol. Exp. Ther., 1966, 153, 134-141.
7. Harris, R.A., Loh, H.H. and Way, E.L. J. Pharmacol. Exp. Ther. 1975, 195, 488-498.
8. Williams, R.J.P. Q. Rev. Chem. Soc., 1970, 24, 331-365.
9. Pressman, B.C. Fed. Proc., 1973, 32, 1698-1703.
10. Chapman, D.B., Louie, G., and Way, E.L., unpublished observations.
11. Weger, P. and Amsler, C. Arch. Exp. Pathol. Pharmakol., 1936, 181: 489-493.
12. Detrick, L.E. and Thienes, C.H. Arch. Int. Pharmacodyn. 1941, 66, 130-137.
13. Kakunaga, T. and Kaneto, H. Abstr. 23rd Int. Longr. Physiol. Sci., Tokyo, 1965, 526.
14. Sanghvi, I.S. and Gerson, S. Life Sci. 1976, 18, 649-654.
15. Harris, R.A., Iwamoto, E.T., Loh, H.H. and Way, E.L. Brain Res. 1975, 100, 221-225.
16. Harris, R.A., Loh, H.H. and Way, E.L. J. Pharmacol. Exp. Ther., 1976, 196, 288-297.
17. Iwamoto, E.T., Harris, R.A., Loh, H.H. and Way, E.L. J. Pharmacol. Exp. Ther., 1978, 206, 46-55.
18. Weiss, G.D. Ann. Rev. Pharmacol., 1974, 14, 343-354.
19. Takata, M., Pickard, W.F., Lettvin, J.Y. and Moore, J.W. J. Gen. Physiol., 1967, 50, 461-471.
20. Kuhar, M.J., Pert, C.B. and Snyder, S.H. Nature, 1973, 245, 447-450.
21. Reddy, S.V.R. and Yaksh, T.L. Neuropharmacol., 1980, 19, 181-185.
22. Henderson, G., Hughes, J. and Kosterlitz, H.W. Nature, 1978, 271, 677-679.
23. Wesche, D.L. and Frederickson, R.C.A. Life Sci., 1979, 24, 1861-1868.
24. Caruso, T.P. and Takemori, A.E. Fed. Proc. 1978, 37(1), 568.
25. Lewis, J.W., Caldecott-Hazard, S., Cannon, J.T. and Liebeskind, J.C. "Neurosecretion and brain peptides: Implications for brain function and neurological disease", Ed. J.B. Martin, Raven Press, New York, 1980.
26. Huidobro-Toro, J.P. and Way, E.L. J. Pharmacol. Exp. Ther., 1979, 211, 50-58.
27. Zieglgänsberger, W. and Bayerl, Brain Res. 115, 111-128.
28. Takemori, A.E. Science, 1961, 133, 1018-1019.
29. Takemori, A.E., J. Pharmacol. Exp. Ther., 1962, 135, 89-93.
30. Elliott, H.W., Kokka, N. and Way, E.L. Proc. Soc. Exp. Biol. Med., 1963, 113, 1049-1052.

31. Nutt, J.G. Fed. Proc., 1968, 27, 753.
32. Opmeer, F.A. and Van Ree, J.M. Europ. J. Pharmacol., 1979, 53, 395-397.
33. Opmeer, F.A. and Van Ree, J.M. J. Pharmacol. Exp. Ther., 1980, 213, 188-195.
34. Hu, J., Hudiobro-Toro, J.P. and Way, E.L. "Endogenous and exogenous opiate agonists and antagonists", Ed. E.L. Way, Pergamon Press, 1979, pp. 263-266.
35. Paton, W.D.M. Br. J. Pharmacol., 1957, 11, 119-127.
36. Huidobro-Toro, J.P., Hu, J. and Way, E. Leong. J. Pharmacol. Exp. Ther., 1981, 218, 84-91.
37. Heimans, R.L.H. Arch. Int. Pharmacodyn., 1975, 215, 13-19.
38. Puig, M.M., Gascon, P. and Musacchio, J.M. J. Pharmacol. Exp. Ther., 1978, 206, 289-302.
39. Bennett, M.R. and Lavidis, N.A. Br. J. Pharmacol., 1980, 69, 185-191.
40. Bennett, M.R. and Florin, T. Br. J. Pharmacol., 1975, 55, 97-104.
41. Illes, P., Ziegelgänsberger, W. and Herz, A. Brain Res. 1980, 191, 511-522.
42. Dingledine, R. and Goldstein, A. J. Pharmacol. Exp. Ther, 1976, 196, 97-106.
43. Williams, J.T. and North, R.A. Brain Res., 197, 165, 57-65.
44. Mudge, A.W., Leeman, S.E. and Fischbach, G.D. Proc. Natl. Acad. Sci. U.S.A., 1979, 76, 526-530.
45. Shikimi, T., Kaneto, H. and Hano, K., Jap. J. Pharmacol., 1967, 17, 135-136.
46. Ross, D.H., Medina, M.A. and Cardenas, H.L. Science 1974, 186, 63-65.
47. Cardenas, H.L. and Ross, D.H. J. Neurochem., 1975, 24, 487-493.
48. Hood, W.F. and Harris, R.A. Biochem. Pharmacol., 1979, 28, 3075-1080.
49. Cardenas, H.L. and Ross, D.H. Br. J. Pharmacol., 1976, 57, 521-526.
50. Yamamoto, H., Harris, R.A., Loh, H.H. and Way, E.L., J. Pharmacol. Exp. Ther., 1978, 205, 255-264.
51. Harris, R.A., Yamamoto, H., Loh, H.H. and Way, E.L. "Opiates and endogenous opioid peptides, Ed. H.W. Kosterlitz, Amsterdam: Elsevier/North Holland Biomed. Press, 1976, pp. 361-368.
52. Harris, R.A., Yamamoto, H., Loh, H.H. and Way, E.L., Life Sci., 1977, 20, 501-506.
53. Ross, D.H. fractions during chronic morphine treatments. Neurochem. Res., 1977, 2, 581-593.
54. Ross, D.H. and Cardenas, H.L. Life Sci., 1977, 20, 1455-1462.

55. Kaku, T., Kaneto, H. and Koida, M. Jap. J. Pharmacol., 1974, 24, 123 (Suppl.).
56. Hoss, W., Okumura, K., Formaniak, M. and Tanaka, R. Life Sci., 1979, 24, 1003-1009.
57. Kakunage, T. Folia Pharmacol. Jpn. 1966, 62, 40-50.
58. Guerrero-Munoz, F., Cerreta, K.V., Guerrero, M.L. and Way, E.L. J. Pharmacol. Exp. Ther., 1979 209, 132-136.
59. Guerrero-Munoz, F., Guerrero, M.L., Way, E.L. and Li, C.H. Science, 1979, 206, 89-91.
60. Guerrero-Munoz, F., Guerrero, M.L. and Way, E.L. J. Pharmacol. Exp. Ther., 1979, 211, 370-374.
61. Blaustein, M.P. and Goldring, J.M. J. Physiol. (London), 1975, 247, 589-615.
62. McCain, H.W., Yamamoto, H. and Way, E.L. Fed. Proc., 1980, 39, 995.
63. Yamamoto, H., McCain, H. and Way, E.L. Fed. Proc., 1980, 39, 759.
64. Hiller, J.M., Pearson, J. and Simon, E.J. Res. Commun. Chem. Pathol. Pharmacol., 1973, 6, 1052-1062.
65. Ross, D.H. "Calcium in drug action", Ed. G.B. Weiss, Plenum Press, New York, 1978, pp. 241-259.
66. End, D., Dewey, W.L. and Carchman, R.A. Fed. Proc., 1978, 37(3), 763.
67. End, D., Warner, W., Dewey, W.L. and Carchman, R.A. Fed. Proc. 1977, 36, 943.
68. Way, E. Leong. "Neurochemical mechanisms of opiates and endorphins" (Adv. Biochem. Pharmacol. Vol. 20), Eds. H.H. Loh and D.H. Ross, Raven Press, New York, 1979.
69. Way, E. Leong. "Opiate receptors and the neurochemical correlates of pain", Ed. Susanna Furst, Pergamon Press, 1980, pp. 153-164.
70. Katz, B. and Miledi, R. Proc. R. Soc. Ser. B, 1965, 161, 496-503.
71. Blaustein, M.P., Johnson, E.M. Jr. and Needleman, P., Proc. Natl. Acad. Sci. U.S.A., 1972, 69, 2237-2240.
72. Schmidt, W.K. and Way, E.L. J. Pharmacol. Exp. Ther. 1980, 212, 22-27.
73. Schmidt, W.K. Ph.D. Thesis, Univ. Calif, San Francisco, 1978, 177 pp.
74. Mulé, S.J. "Narcotic drugs; Biochemical pharmacology", Ed. D.H. Clouet, Plenum Press, New York, 1971, pp. 190-215.
75. Mulé, S.J. Biochem. Pharmacol., 1969, 18: 339-346.
76. Greenberg, S., Diecke, F.P.J. and Long, J.P. J. Pharm. Sci., 1972, 61, 1471-1473.
77. Porcellati, G., Arienti, G., Pirotta, M. and Giorgini, D., J. Neurochem., 1971, 18, 1395-1417.
78. Natsuki, R., Hitzemann, R. and Loh, H.H. Mol. Pharmacol., 1978, 14, 448-453.
79. Putney, J.W., Jr., Weiss, S.J., Van De Walle, C.M. and Haddas, R.A. Nature, 1980, 284, 345-347.

80. Cho, T.M., Law, P.Y., and Loh, H.H. "Neurochemical mechanisms of opiates and endorphins", Eds. H.H. Loh and D.H. Ross, Raven Press, 1979, pp. 69-102.
81. Loh, H.H. and Law, P.Y. Ann. Rev. Pharmacol. Toxicol., 1980, 20, 201-234.
82. Loh, H.H., Shen, F.-H. and Way, E.L. Biochem. Pharmacol., 1969, 18, 2711-2721.
83. Skou, J.C. Biochim. Biophys. Acta, 1957, 23, 394-401.
84. Ray, A.K., Mukherji, M. and Ghosh, J.J. J. Neurochem., 1968, 15, 875-881.
85. Ghosh, S.K. and Ghosh, J.J. J. Neurochem., 1968, 15, 1375-1376.
86. Jain, M.L., Curtis, B.M. and Bakutis, E.V. Res. Commun. Chem. Pathol. Pharmacol., 1974, 7, 229-232.
87. Desaiah, D. and Ho, I.K. Biochem. Pharmacol., 1977, 26, 89-92.
88. Desaiah, D. and Ho, I.K. J. Pharmacol. Exp. Ther., 1979, 208, 80-85.
89. Yamamoto, H., Harris, R.A., Loh, H.H. and Way, E.L., Life Sci., 1977, 20, 1533-1539.
90. Poisner, A.M. and Trifaro, J.M. Mol. Pharmacol., 1967, 3, 561-571.
91. Trifaro, J.M. and Poisner, A.M. Mol. Pharmacol., 1967, 3, 572-580.
92. Kendrick, N.C., Blaustein, M.P., Fried, R.C. and Ratzlaff, R.W. terminals. Nature, 1977, 265: 246-248.
93. Cheung, W.Y. J. Biol. Chem., 1971, 246, 2859-2869.
94. Brostom, C.O., Huang, Y.C., Breckenridge, B.M. and Wolff, D.J. Proc. Natl. Acad. Sci. U.S.A. 1975, 72, 64-68.
95. Cheung, W.Y. Science, 1980, 207, 19-27.
96. Ross, D.H., Cardenas, H.L. and Grady, M.M. Fed. Proc., 1979, 38(3), 253.
97. Levin, R.M., and Weiss, B. J. Pharmacol. Exp. Ther., 1979, 208, 454-459.
98. Bradham, L.S., Holt, D.A. and Sims, M. Biochim. Biophys. Acad, 1970, 201: 250-260.
99. Olson, D.R., Kon, C. and Breckenridge, B.M. Life Sci., 1976, 18, 935-940.
100. Minneman, K.P. and Iversen L.L. "Opiates and endogenous opioid peptides", Ed. H.W. Kosterlitz, Elsevier/North Holland Biomed. Press, Amsterdam, 1976, pp. 361-368.
101. Iversen, L.L. and Minneman, K.P. Trans. Amer. Soc. Neurochem., 1977, 8, 108.
102. Bonnet, K.A. and Gusik, S. Trans. Amer. Soc. Neurochem., 1977, 8, 84.
103. Sharma, S.K., Klee, W.A. and Nirenberg, M. Proc. Natl. Acad. Sci. U.S.A., 1975, 72, 3092-3096.
104. Wilkening, D. Fed. Proc., 1979, 38, 253.
105. Brandt, M., Buchen, C. and Hamprecht, B. cells. J. Neurochem., 1980, 34, 643-651.

106. Maeno, H., Johnson, E.M. and Greengard, P. J. Biol. Chem., 1971, 246, 134-142.
107. DeLorenzo, R.J. Biochem. Biophys. Res. Commun., 1976, 71, 590-597.
108. Clouet, D.H., O'Callaghan, J.P and Williams, N. "Characteristics and Function of Opioids", Eds. J.M. Van Ree and L. Terenius, Elsevier/North Holland Biomed. Press, 1978, pp. 351-352.
109. O'Callaghan, J.P., Williams, N. and Clouet, D.H. J. Pharmacol. Exp. Ther., 1979, 208, 96-105.
110. Clouet, D.H. and Williams, N. "Endogenous and exogenous opiate agonists and antagonists", Ed. E.L. Way, Pergamon Press, 1980, pp. 239-242.
111. Grady, M. and Ross, D. Fed. Proc., 1980, 39, 844.
112. Juskevich, J., O'Callaghan, J. and Lovenberg, W. Fed. Proc., 1979, 38, 253.
113. Shikimi, T., Kaneto, H. and Hano, K. Jap. J. Pharmacol., 1967, 17, 136-137.
114. Sanfacon, G. and Labrecque, G. Psychopharmacol., 1977, 55, 151-156.
115. Sanfacon, G., Houde-Depuis, M., Vanier, R. and Lbrecque, G., J. Neurochem., 1977, 28, 881-884.
116. Jhamandas, K., Sawynok, J. and Sutak, M. Europ. J. Pharmacol., 1978, 49, 309-312.
117. Göthert, M., Pohl, I.-M. and Wehking, E. Naunyn Schmiedeberg's Arch. Pharmacol., 1979, 307, 21-27.
118. Rasmussen, H. and Goodman, D.B.P. Physiol. Rev., 1977, 57, 421-509.
119. Göthert, M. and Wehking, E. Experientia, 1980, 36, 239-241.
120. Down, J.A. and Szerb, J.C. Br. J. Pharmacol., 1980, 68, 47-55.
121. Szerb, J.C. Naunyn Schmideberg's Arch. Pharmacol., 1980, 311, 119-127.
122. Cowie, A.L., Kosterlitz, H.W. and Waterfield, A.A. Br. J. Pharmacol., 1978, 64, 565-580.
123. Kilbinger, H. and Wagner, P. Naunyn Schmiedeberg's Arch. Pharmacol., 1978, 287, 47-60.
124. Sawynok, J. and Jhamandas, K. Can. J. Physiol. Pharmacol., 1979, 57, 853-859.
125. Westfall, T.C. Physiol. Rev., 1977, 57, 659-728.
126. Hollins, C. and Stone, T.W. Br. J. Pharmacol. 1980, 69, 107-112.

RECEIVED June 14, 1982.

8

Calcium Antagonism and Antiepileptic Drugs

JAMES A. FERRENDELLI

Washington University Medical School, Division of Clinical Neuropharmacology, Department of Neurology and Neurological Surgery and Department of Pharmacology, St. Louis, MO 63110

The mechanisms of action of most antiepileptic drugs presently in use are poorly understood. However, it is generally agreed that their anticonvulsant effect are probably due to some action on neurotransmission or a direct action on membrane function, particularly ionic conductances, or both in CNS. We have attempted to evaluate the influence of antiepileptic drugs on calcium conductance in brain tissue. Studies of the effects of phenytoin on Ca permeability in isolated nerve terminals (synaptosomes) indicate that this drug inhibits depolarization induced Ca uptake. It appears to selectively reduce sodium permeability and, as a result of this, diminish the degree of membrane depolarization, thus, indirectly, reducing stimulus-coupled Ca accumulation. Phenytoin also inhibits Ca uptake by a sodium-independent process, and we suggest that this may be a result of its presence in cellular membranes and physically distorting their organization. Carbamazepine and lidocaine have effects on Ca conductances similar to those of phenytoin, but other antiepileptic drugs are relatively ineffective. These data, in conjunction with other reported results, lead to the conclusion that the anticonvulsant actions of phenytoin, carbamazepine and lidocaine are partially due to actions on ionic (Na and Ca) permeability in nervous tissue. The relationship between this mechanism and the selective clinical effects of antiepileptic drugs is discussed.

Primary, commonly used antiepileptics include phenytoin, phenobarbital, primidone, carbamazepine, ethosuximide and valproic acid. In addition, two benzodiazepines, diazepam and

0097-6156/82/0201-0143$06.00/0

clonazepam, are used extensively to treat seizure disorders. These compounds represent six different chemical classes and thus may be expected to have diverse actions. However, the mechanisms of action of antiepileptic drugs are poorly understood at the present time. Presently, available information suggests that augmentation of inhibitory neurotransmission and alteration of ionic conductances in excitable membranes may be important for the antiepileptic effects of many available drugs. The relationship betwen anticonvulsant drugs and calcium is of particular interest. Studies in our laboratory have suggested that some antiepileptic drugs affect calcium conductances in nervous tissue and may also affect other ionic conductances (1-3). The present report reviews these studies and a hypothesis explaining the role calcium antagonism in antiepileptic drug mechanism is presented.

Experimental Methods

In the experiments described in this report calcium conductance was measured in nerve terminals (synaptosomes) isolated from rat cerebral cortex. Calcium conductances in these subcellular particles have been characterized and described extensively by Blaustein and colleagues and it appears that synaptosomes possess many of the properties of *in situ* nerve terminals (4-7).

Briefly, synaptosomes were prepared (8) and calcium uptake was measured in the following manner. Four rats were killed by asphyxiation and their brains were rapidly removed and placed in ice-cold 0.32 M sucrose. The forebrains, excluding olfactory bulbs, were homogenized in 5 volumes of sucrose in a teflon-glass homogenizer. After centrifugation to remove nuclei and debris, the synaptosomes were isolated by differential centrifugation in a discontinuous sucrose gradient. Ice-cold buffer, which was identical to control buffer (see below) except that it contained only 0.02 mM $CaCl_2$, was slowly added to the 0.8 M sucrose fraction enriched with synaptosomes to adjust the sucrose concentration to 0.32 M. Synaptosomes were recovered by centrifugation and suspended in control buffer containing 132 mM NaCl, 5 mM KCl, 1.2 mM each of $CaCl_2$, $MgCl_2$ and NaH_2PO_4, 2 mM tris, and 10 mM glucose, pH 7.4, to give a final concentration of 1.2-1.6 mg prot/ml.

Aliquots (0.5 ml) of synaptosomal suspension were preincubated for 25 min at 30° C. When effects of drugs were examined, they were added to the synaptosomal suspension at the beginning of the preincubation period at various concentrations. Following preincubation ^{45}Ca uptake was initiated by the addition of 0.5 ml of one of the following solutions, each containing 0.5 μCi $^{45}CaCl_2$/ml: (1) control buffer; (2)

control buffer containing various concentartions of veratridine; (3) a buffer identical to control buffer except that it contained 17 mM NaCl and 120 mM KCl. In experiments where effects of drugs were examined, they were included in these buffers at designated concentrations. In some experiments (see results), $MnCl_2$ or tetrodotoxin was included in the buffers. Following incubation in the presence at ^{45}Ca for various times as indicated in the results section, uptake was terminated by the rapid addition of 6 ml of ice-cold stop buffer containing 132 mM NaCl, 5 mM KCl, and 3 mM EGTA adjusted to pH 7.4 with tris base. Immediately following this the suspension was filtered through glass fiber filters (GF/A Whatman, Inc.). The filters were washed with an additional stop buffer and then placed in glass vials containing 10 ml of scintillation cocktail (Scintiverse, Fisher Scientific Co.) and counted in a liquid scintillation counter. Results were expressed as stimulated calcium uptake, which is the accumulation of Ca in the presence of veratridine or high K minus the uptake in control buffer.

Characteristics of Calcium Uptake in Synaptosomes

Addition of ^{45}Ca to synaptosomes incubated in control buffer (132 mM Na, 5 mM K) results in a rapid uptake of the isotope during the first five seconds, a slower accumulation for the next 25 seconds, and very little or no additional uptake thereafter. Increasing the [K] in the incubation media markedly augments Ca uptake. This stimulatory effect of K seems to occur immediately and is essentially complete within the first 30 sec. Veratridine also increases synaptosomal Ca uptake, but its onset of action is slower than that of K and requires 3 min or longer for completion. The maximum effect of veratridine is similar to that of K, however, and the stimulatory effect of both K and veratridine on Ca uptake are concentration-dependent.

Tetrodotoxin, at concentrations of 10^{-7} M or greater, blocks the action of veratridine 90% but has no effect on the action of K, even at concentrations up to 10^{-6} M (Figure 1). In contrast, Mn inhibits Ca uptake stimulated by veratridine and K in similar fashions, with an apparent ID_{50} of 1 mM for both and essentially complete blockade at 10 mM (Figure 1).

Apparently K and veratridine have different mechanisms for stimulating calcium uptake. High concentrations of extracellular K reduce the normally large gradient between intra- and extracellular K levels and cause membrane depolarization. In contrast, veratridine prevents inactivation of Na channels, thus allowing excessive accumulation of Na intracellularly, and this, in turn, results in membrane depolarization. In either case Ca uptake is stimulated by the depolarization. Since the effect of veratridine is Na-dependent, it is inhibited by tetrodotoxin, a specific blocker of Na channels (9). In contrast, the action of K is Na-independent and is unaffected by tetrodotoxin. Obviously, Mn, which acts primarily by occluding Ca channels (10), has an equivalent effect on K- and veratridine-induced Ca uptake.

Effect of Phenytoin and Other Hydantoins

Phenytoin also affects calcium uptake in synaptosomes. Similar to Mn, it inhibits the action of both veratridine and K, but like tetrodotoxin it is much more effective against veratridine (Figure 2). Stimulated Ca uptake produced at all concentrations of K between 15 and 64 mM is inhibited 20% by 0.1 mM phenytoin, and this is unaffected by tetrodotoxin. Examination of the dose-related effect of phenytoin on stimulated Ca uptake produced by 23 and 64 mM K reveals identical proportional inhibition of both. Thus, inhibition of K-induced Ca uptake by phenytoin is independent of K concentration. In contrast, there appears to be a competitive interaction between veratridine and phenytoin. Phenytoin has a much greater inhibitory effect, proportionally, on Ca uptake stimulated by low concentrations of veratridine than that produced by higher concentrations. For example, 35 μM phenytoin caused 50% inhibition of Ca uptake produced by 5 μM veratridine, but ~ 225 μM phenytoin is required for 50% inhibition of the Ca uptake produced by 100 μM veratridine.

Thus, phenytoin appears to inhibit K-induced and veratridine-induced calcium uptake by different mechanisms. Present data demonstrate that phenytoin has an action similar, but not identical, to that of tetrodotoxin, as well as an action like, but also not identical, to that of Mn. Other investigators have reported that phenytoin blocks Na channels in lobster nerve and squid giant axon (11-13) and this is not a Ca-dependent process. The present data demonstrate that phenytoin is able to block K-stimulated Ca uptake, and this is unaffected by tetrodotoxin, indicating that this effect is independent of Na. We think that all of the data together suggest that phenytoin can inhibit both Na and Ca conductances in nervous tissue and that these are separate and independent processes. The present results also indicate that Na conductance may be more sensitive to the inhibitory action of phenytoin, but more research will be necessary to prove this contention.

The effects of phenytoin on K- and veratridine-induced Ca uptake have been compared with those of other hydantoins (Table 1). These include the major *in vivo* metabolite of phenytoin, hydroxyphenyl-phenylhydantoin (HPPH); mephenytoin, another antiepileptic hydantoin; its major metabolite, 5-ethyl-5-

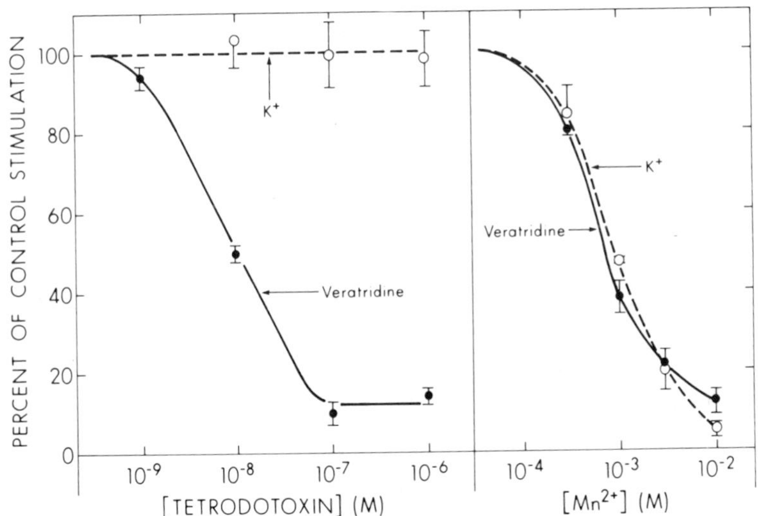

Figure. 1. Effect of tetrodotoxin (left) and Mn^{2+} (right) on stimulated Ca uptake by rat brain synaptosomes. Uptake of ^{45}Ca was stimulated by exposing synaptosomal suspensions to 5 μM veratridine (●) for 5 min or to 64 mM K^+ (○) for 15 s in the absence or presence of the indicated concentrations of tetrodotoxin or $MnCl_2$. Values are expressed as percentage of control (absence of tetrodotoxin and Mn) stimulation (4.1 ± 0.7 μmol Ca accumulated/g prot). Each point and vertical line represents the mean and SEM, respectively, of triplicate samples.

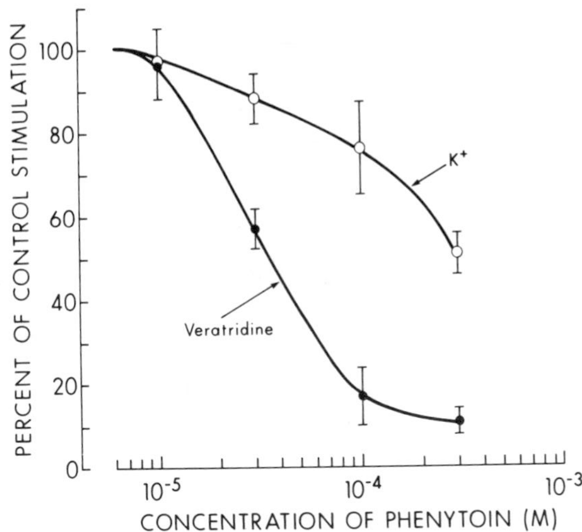

Figure 2. Effect of phenytoin concentration on K- and veratridine-stimulated Ca uptake in rat brain synaptosomes. Synaptosomal uptake of ^{45}Ca was stimulated by exposure of the tissue to 64 mM K^+ (○) for 15 s or 5 μM veratridine (●) for 5 min. In the absence of phenytoin control, stimulation was 4.7 ± 0.3 and 6.1 ± 0.8 μmol Ca/g prot for K^+ and veratridine, respectively. Each symbol and vertical line represents the mean and SEM, respectively, of 6 samples.

phenylhydantoin; and hydantoin itself. All compounds except hydantoin inhibit both K- and veratridine-induced Ca uptake at concentrations less than 1 mM (the highest dose tested). However, HPPH, mephenytoin and its demethylated derivative are all less potent than phenytoin by an order of magnitude. Similar to phenytoin, the other effective hydantoins are more active against veratridine than against K-stimulated Ca accumulation, although none exhibit as much selectivity as phenytoin. Thus, the inhibitory action of phenytoin on ionic permeability is significantly altered by modification of its molecular structure. The addition of a hydroxyl group to one phenyl ring or replacement of a phenyl group with an alkyl group diminishes potency and seems to decrease the relatively greater inhibitory action against Na-dependent Ca uptake. These changes in effects may simply be due to lower lipid solubility of the other hydantoin derivatives. Regardless, the results suggest that phenytoin has actions which are quantitatively, and perhaps qualitatively, different from those of other anticonvulsant hydantoins.

Table 1

Inhibitory Effects of Hydantoins on K- and Veratridine-Stimulated Calcium Uptake in Synaptosomes

Drug	Drug Concentration (μm) Producing 50% Inhibition of Ca Uptake	
	Veratridine	K
Phenytoin	35	350
Hydroxyphenyl-phenylhydantoin	210	900
Mephenytoin	800	1200
5-Ethyl-5-phenyl-hydantoin	450	1600
Hydantoin	10,000	10,000

Effect of Other Antiepileptic Drugs

We have also tested several other anticonvulsant drugs and found that carbamazepine, phenobarbital, lidocaine, and diazepam inhibit stimulated Ca uptake in synaptosomes (Table 2). Ethosuximide and valproic acid, however, have no effect exept at a concentration of 10 mM, and even at this extremely high concentration they produce only very little inhibition. Phenobarbital is clearly less potent than phenytoin by an order of magnitude or greater , but carbamazepine, lidocaine and diazepam are nearly equipotent with phenytoin. Similar to phenytoin, carbamazepine and lidocaine blocked veratridine-stimulated Ca uptake much better than K-stimulated uptake. Phenobarbital and diazepam have little or no differential effect on the two depolarizing agents, however. Thus, with the exception of ethosuximide and valproic acid, all antiepileptic drugs tested are effective. However, phenobarbital is considerably less potent, and both phenobarbital and diazepam do not appear to have any slective effect against veratridine-iduced Ca uptake.

Table 2

Inhibitory Effects of Some Anticonvulsant Drugs on K- and Veratridine Stimulated Calcium Uptake in Synaptosomes

Drug	Drug Concentration (μm) Producing 50% Inhibition of Ca Uptake	
	Veratridine	K
Phenytoin	35	350
Carbamazepein	100	1800
Lidocaine	10	> 1000
Phenobarbital	1350	4500
Diazepam	65	100
Ethosuximide	>10,000	>10,000
Valproic Acid	>10,000	>10,000

Thus, only carbamazepine and lidocaine have actions very similar to those of phenytoin. Lidocaine, as well as most other local anesthetics, inhibits Na conductance in nerve fibers (14, 15). There are also limited data indicating that carbamazepine is capable of blocking Na permeability in *Myxicola* giant axons (16). These data, in conjunction with the present results, suggest that carbamazepine and lidocaine inhibit both Na and Ca conductances in nervous tissues similar to phenytoin.

The mechanisms responsible for inhibition of ionic conductances exhibitedby the several drugs described above are poorly understood at present. It has been proposed that local anesthetics (e.g., lidocaine) selectively block Na conductance by combining with specific receptor sites in or near the Na channel (14,17). Possibly phenytoin and carbamazepine also act at sites on or near Na channels, and these two drugs may have specific and selective action on Na conductance, perhaps similar to that of lidocaine. Whether these drugs and others examined in this study block Ca conductance by a direct or indirect process is difficult to ascertain. We are impressed, however, with the fact that the more lipid soluble agents (e.g., diazepam) are more effective inhibitors of Na-independent Ca uptake than the more water soluble compounds (e.g., phenobarbital). Perhaps Na-independent inhibition of Ca conductance results from a drug entering cellular membranes in sufficient quantity to physically distort their organization and by this means modify the structure and function of ionic channels. We suggest that this mechanism may best explain inhibition of K-induced Ca uptake observed with phenytoin, lidocaine and carbamazepine and may also explain the actions of diazepam and phenobarbital on both K- and veratridine-induced Ca uptake.

Role of Calcium and Sodium Antigonism in Anticonvulsant Mechanisms

An issue worthy of consideration is the question of whether the antiepileptic or other clinical effects of any of the drugs examined here are wholly or partially a result of their inhibition of Ca and/or Na movement across cellular membranes. Although there are several factors that could be considered when attempting to answer this question, one of the more important is drug concentration. Obviously, if an action of a drug is related to its clinical effect, both would be expected to occur at the same or nearly the same drug concentration. Therapeutic concentrations of antiepileptic drugs (18) are listed in Table 3. Only phenytoin, carbamazepine, and perhaps lidocaine inhibit veratridine-induced Ca accumulation at concentrations which approximate their "therapeutic levels." K-induced Ca uptake is also inhibited a small amount (10-20%) by "therapeutic levels" of phenytoin. All of the other actions observed in this study occur at drug concentrations which are 1-2 or more orders of magnitude higher than those usually achieved clinically. Thus, we conclude that inhibition of Na and/or Ca uptake could be a mechanism of action underlying the clinical effects of phenytoin, carbamazepine and lidocaine, but this process probably is not involved in the clinical effects of the other drugs examined in this study.

Table 3

Clinically Effective Blood Levels of Some Antiepileptic Drugs

Drugs	Therapeutic Blood Level (μm)
Phenytoin	40 - 80
Carbamazepine	20 - 50
Phenobarbital	60 - 160
Diazepam	1
Ethosuximide	350 - 700
Valproid Acoid	350 - 650
Lidocaine	10 - 20

Furthermore, we think that one may also conclude that the clinical effects of phenytoin, carbamazepine and lidocaine are most likely due to their inhibitory action on Na uptake rather than a primary action on Ca conductance. This is based on the finding that much lower concentrations of all three drugs are needed to inhibit veratridine-stimulated (i.e., Na-dependent) Ca uptake than are required to inhibit K-stimulated (Na-independent) Ca uptake. This does not necessarily imply that an alteration of Ca conductances is not involved in the clinical effects of the above drugs. In the CNS, in situ, Ca uptake is stimulated by cellular depolarization which results from augmented Na influx. Obviously, impairment of Na influx by a drug would secondarily diminish Ca uptake. Also, at least in the case of phenytoin, direct inhibition of Ca uptake may have a role in its clinical effect.

Acknowledgment

Supported, in part, by USPHS Grant NS-14834

Literature Cited

1. Sohn, R. S.; Ferrendelli, J. A. J. Pharmacol. Exp. Ther. 1973, 185, 272-275.
2. Sohn, R. S.; Ferrendelli, J. A. Arch. Neurol. 1976, 33, 626-629.
3. Ferrendelli, J. A.; Daniels-McQueen, S. J. Pharmacol. Exp. Ther. 1982, 220, 29-34.
4. Blaustein, M. P. J. Physiol. 1975, 247, 627-655.
5. Blaustein, M. P.; Ector, A. C. Mol. Pharmacol. 1975, 11, 369-378.
6. Blaustein, M. P.; Ector, A.C. Biochim. Biophys. 1976, 419, 295-308.
7. Blaustein, M. P.; Goldring, J. M. J. Physiol. 1975, 247, 589-615.
8. Krueger, B. K.; Ratzlaff, R. W.; Strichartz, G. R.; Blaustein, M. P. J. Memb. Biol. 1979, 50, 287-310.
9. Narahashi, T. Physiol. Rev. 1974, 54, 813-888.
10. Nachshen, D. A.; Blaustein, M. P. J. Gen. Physiol. 1980, 76, 709-728.,
11. Lipicky, R. J.; Gilbert, D. L.; Stillman, I. M. Proc. Natl. Acad. Sci. USA 1972, 69, 1758-1760.
12. Pincus, J. H. Arch. Neurol. 1972, 26, 4-10.
13. Perry, J. G.; McKinney, L.; Deweer, P. Nature 1978, 272, 271-273.
14. Strichartz, G. R. J. Gen. Physiol. 1973, 62, 37-57.
15. Ritchie, J. M. Br. J. Pharmacol. 1975, 74, 191-198.
16. Schant, C. L.; Davis, F. A.; Marder, V. J. Pharmacol. Expt. Ther. 1974, 189, 538-543.
17. Hille, B. J. Gen. Physiol. 1977, 69, 497-515.
18. Kutt, H.; McDowell, F. "Clinical Neuropathology"; Churchill Livingstone: New York, 1979, pp. 12-53.

RECEIVED June 1, 1982.

9

Relationship of Cyclic Nucleotides and Calcium in Platelet Function

G. C. LE BRETON, N. E. OWEN, and H. FEINBERG

University of Illinois College of Medicine, Department of Pharmacology, Chicago, IL 60612

Platelets are intimately involved in the prevention of blood loss upon vascular damage as well as in the genesis of certain forms of cardiovascular disease. Although the precise mechanisms involved in the control of platelet function are unknown, Ca^{2+} and adenosine 3'5'-cyclic monophosphate (cAMP) appear to modulate the levels of platelet reactivity. In this regard, platelet activation is generally associated with an increase in the concentration of cytosolic Ca^{2+}. Thus, an elevation in the levels of intraplatelet Ca^{2+}, whether as a consequence of increased platelet membrane permeability to Ca^{2+} or release of internal Ca^{2+}, is linked to the processes of shape change, aggregation and secretion. On the other hand, agents which stimulate adenylate cyclase activity, e.g., prostacyclin or prostaglandin E_1, are potent inhibitors of platelet function. It has been suggested that cAMP inhibits platelet activation by reducing the availability of cytosolic Ca^{2+}. In this connection, it has been shown that cAMP promotes the uptake of Ca^{2+} into an isolated platelet membrane fraction resembling the sarcoplasmic reticulum of muscle. Furthermore, in intact platelets, increases in cAMP are directly related to enhanced intraplatelet Ca^{2+} sequestration and reduced Ca^{2+} mobilization in response to various agonists. This ability of cAMP to modulate platelet Ca^{2+} levels may be related to the phosphorylation of specific platelet proteins involved in a cAMP-dependent Ca^{2+} pump.

Blood platelets circulate in the vascular system as discrete discoid-shaped cells approximately 2–3 µm in diameter. Although anucleated, platelets contain mitochondria, glycogen particles

0097-6156/82/0201-0153$06.50/0

and various types of storage organelles. Osmophilic granules called dense bodies are the storage sites for adenosine diphosphate (ADP), adenosine triphosphate (ATP), serotonin (5-HT) and Ca^{2+} (1). Various proteins and platelet factor (4) (2) are stored in another population of granules designated alpha granules. In addition, platelets possess a bundle of microtubules which course around the circumference of the cell just beneath the plasma membrane (3). It has been suggested that these microtubules function to maintain the discoid shape of the unstimulated platelet (4) as well as to facilitate the secretion of the platelet granules during cellular activation (5). Platelets also possess two internal membranous structures; one is termed the external canalicular system which has been shown to be an invagination of the outer plasma membrane (6). The other is termed the dense tubular system which is believed to function in a manner analogous to the sarcoplasmic reticulum of muscle in regulating cytosolic Ca^{2+} levels (7).

When isolated platelets are exposed to various physiological agonists, e.g. ADP, they are triggered to undergo a sequence of morphological and functional alterations. One of the first events involves a transition from smooth discoid-shaped cells to cells which are more spherical in shape with blebs or pseudopodia projecting from their surface. The platelet shape change response is followed by a change in the adhesive properties of the external platelet membrane which facilitates platelet-platelet adhesion. Under these conditions the platelets begin to clump together or aggregate.

A separate phenomenon, described by Grette (8) as the release reaction, is the secretion of granular constituents from the alpha granules and the platelet dense bodies. This process appears to be an amplification mechanism whereby the released substances, e.g. ADP, serotonin, Ca^{2+}, etc., promote surrounding platelets to undergo further aggregation. This is presumably the mechanism of platelet plug formation which acts to prevent the loss of blood from a damaged vessel *in vivo* (9).

Under *in vitro* conditions the platelet response of shape change followed by aggregation followed by secretion appears to be a coordinated sequence of a single activating event (10). On the other hand, aggregation need not be preceded by shape change (e.g. as after addition of epinephrine) and secretion can occur as the initial event. Thus shape change, aggregation and secretion are independent processes; however, they appear to be controlled by a common intracellular messenger, i.e. ionic calcium.

Calcium activation of platelet function

There is compelling evidence to suggest that Ca^{2+} serves as a common mediator of platelet functional change. Thus, addition of the divalent cation ionophore, A23187, to a platelet suspension results in shape change, aggregation and secretion (11-17). Since A23187 can induce a net uptake of ^{45}Ca in many cell types (11), it was suggested that platelet activation is triggered by the Ca^{2+} brought in by the ionophore. However, A23187-stimulation of both platelet shape change and secretion also occurs in the absence of external Ca^{2+} (11,12,13,17). Therefore, the Ca^{2+} involved in this process of activation must be derived from an intraplatelet source. On this basis it was proposed that platelets maintain an internally releaseable pool of Ca^{2+}, which when mobilized, serves to initiate platelet functional change. One possible source for this releaseable Ca^{2+} is the platelet dense tubular system (DTS) (7). Thus, a stimulus which causes the release of Ca^{2+} from this membrane system would ultimately lead to an elevation of cytosolic Ca^{2+} levels.

Although this model was consistent with results obtained using A23187, which is known to alter intracellular cation gradients (18,19), a similar mechanism of action by physiological platelet agonists, e.g. ADP, epinephrine, thrombin, thromboxane A_2, etc. had not been demonstrated.

ADP-stimulated Ca^{2+} mobilization. On this basis, we investigated an increase in intracellular Ca^{2+} as a possible mechanism of platelet activation induced by ADP. In these studies it was first demonstrated that ADP-stimulated shape change and aggregation is not associated with a transmembrane Ca^{2+} influx (20). This observation was subsequently confirmed by Massini and Luscher (21). Based on these findings it was concluded that an elevation in cytosolic Ca^{2+} in response to ADP must be derived from intra- platelet storage sites.

In order to pursue this question we developed a technique which permitted the measurement of intracellular Ca^{2+} movements in intact platelets (22). This was accomplished through the use of a Ca^{2+}-sensitive fluorescent probe, chlortetracycline (CTC), and a photon counting microspectrofluorometer.

CTC had been previously used as a means of monitoring changes in intracellular Ca^{2+} binding in many cell systems including sarcoplasmic reticulum (23), red blood cells (24) and mitochondria (25). CTC is known to form a highly fluorescent pH insensitive (pH 6.0-8.5) adduct when chelated with divalent cations bound to biological membranes (26). Even though CTC chelates both Mg^{2+} and Ca^{2+}, it was previously reported using red blood cells (24) and confirmed by our measurements in platelets (22) that the observed fluorescence stems almost entirely from the Ca-CTC adduct. The characteristic which makes CTC a valuable intracellular Ca^{2+} probe is that the fluorescence intensity of the Ca-CTC complex markedly decreases when the Ca^{2+} is released from the membrane to a more polar environment, e.g. the cytosol (27). Consequently, relative changes in cellular fluorescence can be used as an index of transient alterations in the availability of intraplatelet ionic calcium.

Using this procedure we demonstrated that the addition of ADP (1 μM) to platelet rich plasma resulted in a mobilization of platelet Ca^{2+} (as evidenced by a decrease in fluorescence) which occurred coincident with a change in platelet shape (Figure 1). In order to eliminate the possibility that the observed change in fluorescence represented alterations in Ca^{2+} binding to the platelet external membrane, the external Ca^{2+} pool was eliminated by prior addition of EGTA (3 mM).

Furthermore, when the platelets were pretreated with ATP, which acts as a specific ADP antagonist (28), platelet shape change and the change in Ca-CTC fluorescence were both inhibited. It can also be seen that stimulation of platelet shape change by the divalent cation ionophore, A23187, was associated with a similar redistribution of intraplatelet Ca^{2+}. These findings provided evidence that platelet activation induced by ADP is mediated through an internal release of membrane bound Ca^{2+}.

This notion was supported by additional studies in which platelet water was totally exchanged with deuterium oxide (D_2O) (29). D_2O had been previously shown to block the release of Ca^{2+} from the sarcoplasmic reticulum of barnacle muscle and inhibit muscular contraction (30). Since the platelet DTS is thought to function in a manner analogous to the sarcoplasmic reticulum in muscle, we examined the possibility that D_2O would have a similar inhibitory effect in platelets. This was indeed found to be the case, i.e., D_2O inhibited both shape change and aggregation stimulated by ADP (29). Other studies have suggested that D_2O treatment (20-60%) can result in enhanced aggregation (31). In these experiments, however, aggregation was induced with the divalent cation ionophore A23187, rather than ADP. Since it is known that D_2O will also facilitate Ca^{2+}-stimulated muscle contraction (32,33) the effects of D_2O on ionophore-induced platelet activation would be considerably different from its effects on ADP-induced platelet activation. Thus, with A23187, D_2O would augment the effects of available Ca^{2+} whereas with ADP, D_2O would inhibit Ca^{2+} release.

Epinephrine-stimulated Ca^{2+} influx. Our results with D_2O also provided evidence that a different platelet agonist, i.e. epinephrine, does not act through the same mechanism of Ca^{2+} mobilization as does ADP. In this regard, D_2O was not found to be effective in blocking epinephrine-induced platelet activation (29). Therefore, we investigated the possibility that epinephrine, in contrast to ADP, increases platelet membrane permeability to Ca^{2+}. Using trace amounts of ^{45}Ca it was demonstrated that epinephrine does in fact induce a transmembrane Ca^{2+} influx which is concomitant with the

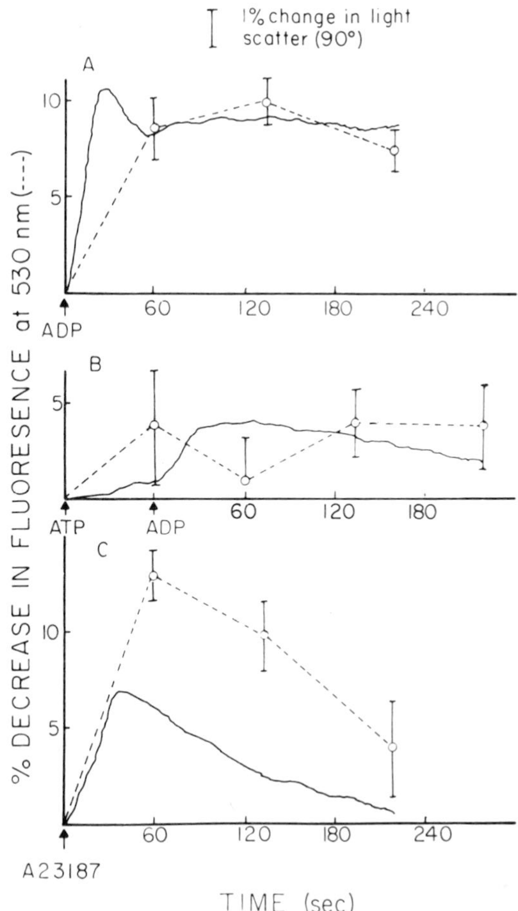

Figure 1. Relationship between light scattering (——) and fluorescence (– – –) of platelets containing chlortetracycline. (Reproduced, with permission, from Ref. 22. Copyright 1976, Biochemical and Biophysical Research Communications.)

Human platelets were incubated at 25°C with 50 μM CTC for 40 min. The plasma was supplemented with 3 mM EGTA to prevent aggregation. Platelets from a portion of the plasma were then pelleted through silicone oil, and fluorescence was determined in a microspectrofluorometer (400 nm excitation: 530 nm emission). A: ADP (1 μM) was added to the plasma, and samples were pelleted and analyzed for fluorescence at 60, 130, and 220 s; each time point represents the mean ± SEM of 25 samples from three different blood donors. Platelet shape change in response to ADP (1 μM) was separately determined by measuring the intensity of scattered light at 90°C. The change in platelet shape represents a typical platelet response in the presence of 50 μM CTC. B: Inhibition of ADP induced shape change and platelet fluorescence change by treatment with ATP (10 μM). ADP (1 μM) was added 60 s after ATP. The fluorescence values at each time point represent the mean ± SEM of 16 samples from two different blood donors. C: Relationship between platelet shape change induced by A23187 (1 μM) and fluorescence change. The fluorescence values at each point represent the mean ± SEM of eight samples from two blood donors.

aggregation response (34) (Figure 2). This ^{45}Ca uptake was dependent upon the dose of epinephrine employed (10–0.1 μM) and was not stimulated by d-epinephrine which is biologically inactive (34). Evidence that a change in platelet membrane permeability to Ca^{2+} was specifically mediated through a receptor interaction was provided by the finding that the α-receptor antagonist, phentolamine, blocked ^{45}Ca uptake as well as platelet aggregation (34). Finally, pretreatment of the platelets with verapamil, which selectively blocks Ca^{2+} channels in cardiac cells, inhibited both epinephrine-stimulated ^{45}Ca uptake (Figure 3) and platelet aggregation but did not inhibit aggregation stimulated by ADP (34). The inhibition by verapamil could in turn be reduced by external Ca^{2+} supplementation.

These findings provided evidence that the mechanism by which epinephrine stimulates platelet functional change is through enhanced membrane permeability to calcium. This effect of epinephrine appears to be in marked contrast to the mechanism of ADP-stimulation which is presumably mediated through internal Ca^{2+} release.

ADP-stimulated Na^+ influx. Although the nature of the ADP interaction with platelet receptors to induce calcium redistribution is unknown, it is clear that ADP does not penetrate the membrane (35). It is possible that the ADP-platelet interaction involves the flux of other ions as part of the stimulus-transfer pathway. In this respect, platelets, like other cells, maintain low intracellular Na^+ levels and high K^+ levels (36); presumably, as a consequence of an Na^+ efflux pump. We therefore investigated the possibility that ADP-induced platelet activation is associated with an altered flux of Na^+.

ADP added to resting platelets induced a sudden rise in ^{22}Na uptake, an increase in platelet Na^+, and a decrease in K^+ (Figure 4) (37). Earlier studies had shown that platelet water volume does not increase after ADP-induced aggregation (38), and using ^{36}Cl, we found no increase in ^{36}Cl uptake (37). An uptake of Na^+ was not seen after GDP addition. Furthermore, the uptake of ^{22}Na occurred simultaneously with the onset of shape change and aggregation (37,39,40). On the other hand, epinephrine-induced aggregation, shown earlier to induce $^{45}Ca^{2+}$ uptake, had no effect to induce ^{22}Na uptake (Figure 5) (39,40). Consequently it appears that cation influx is agonist specific, at least with respect to activation by ADP or by epinephrine.

Whether the uptake of Na^+ associated with ADP-induced shape change and aggregation is the primary stimulus-transfer pathway of activation has not been established. Platelets maintain a membrane potential (negative inside) (41), and activation may ensue in association with a change in membrane potential and an increase in permeability to Na^+. A sudden

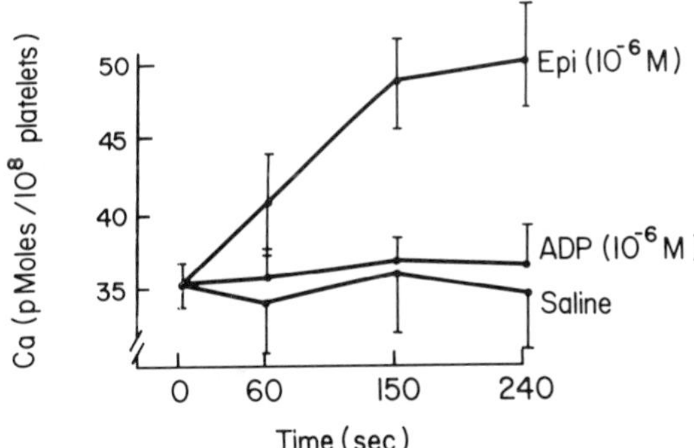

Figure 2. Ca^{2+} uptake associated with primary aggregation induced by epinephrine or ADP. (Reproduced, with permission, from Ref. 34. Copyright 1980, The American Physiological Society.)

Aspirin (1 mg/mL) or indomethacin (20 μM) treated human blood platelets were isolated by albumin density gradient centrifugation and resuspended in Tyrode buffer (pH 7.4) in which the ionic calcium concentration had been adjusted to 100 μM with $CaCl_2$. Thirty min after $CaCl_2$ addition, the platelet suspension was supplemented with fibrinogen (0.1 mg/mL), ^{125}I–human serum albumin, and ^{45}Ca. Three min later, aggregation was induced by epinephrine (1 μM), or ADP (1 μM), and 1 mL aliquots of the platelet suspension were withdrawn at 60, 150, and 240 s following the addition of aggregating agent. The platelets were then analyzed for Ca^{2+} uptake. Each point represents the mean ± SEM of 16 samples from four blood donors.

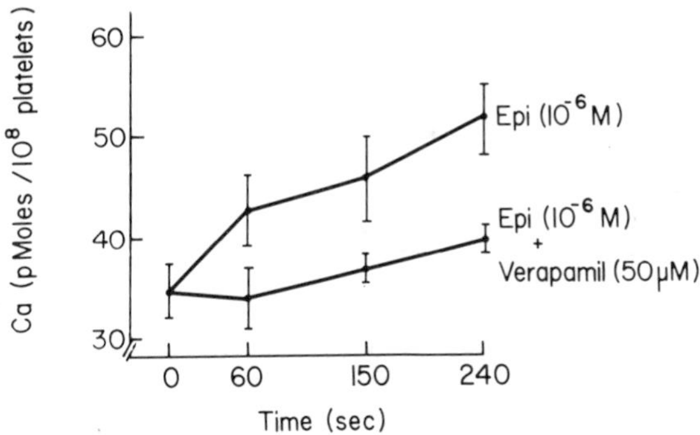

Figure 3. Inhibition of epinephrine-induced Ca^{2+} uptake by verapamil in resuspended platelets. (Reproduced, with permission, from Ref. 34. Copyright 1980, The American Physiological Society.)

Platelets were isolated and Ca^{2+} uptake was determined (conditions were as stated in Figure 2). Resuspended platelets were incubated with verapamil (50 μM) for 1 min before addition of epinephrine (1 μM). Each point represents mean ± SEM of 28 samples from seven separate blood donors.

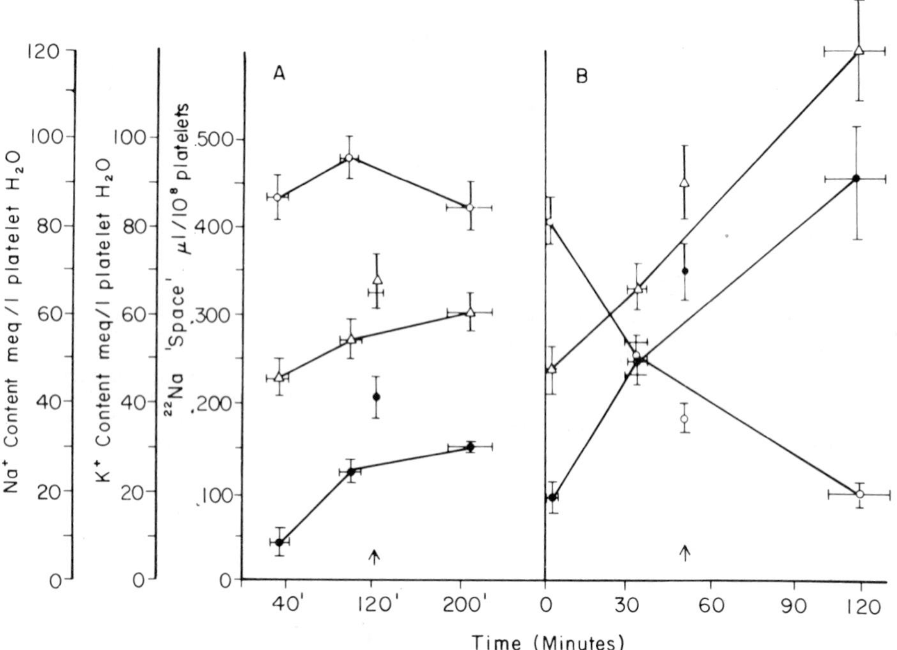

Figure 4. Effect of ADP on platelet Na^+ and K^+ and $^{22}Na^+$ 'space.' Platelet rich plasma was incubated at 37°C in the absence (panel A) or presence (panel B) of ouabain (1 µM). The arrow indicates the time of addition of ADP to an aliquot of the sample. Key: ●, $^{22}Na^+$ 'space'; △, Na^+ content; and ○, K^+ content of the unstimulated platelets. The separate symbols aligned with the arrow depict the same measurements obtained 60 s after the addition of ADP. Samples were taken at slightly different times in four experiments; the time ranges are shown by the horizontal bars. Each point represents the mean ± SEM. (Reproduced, with permission, from Ref. 37. Copyright 1977, Elsevier Biomedical Press B.V.)

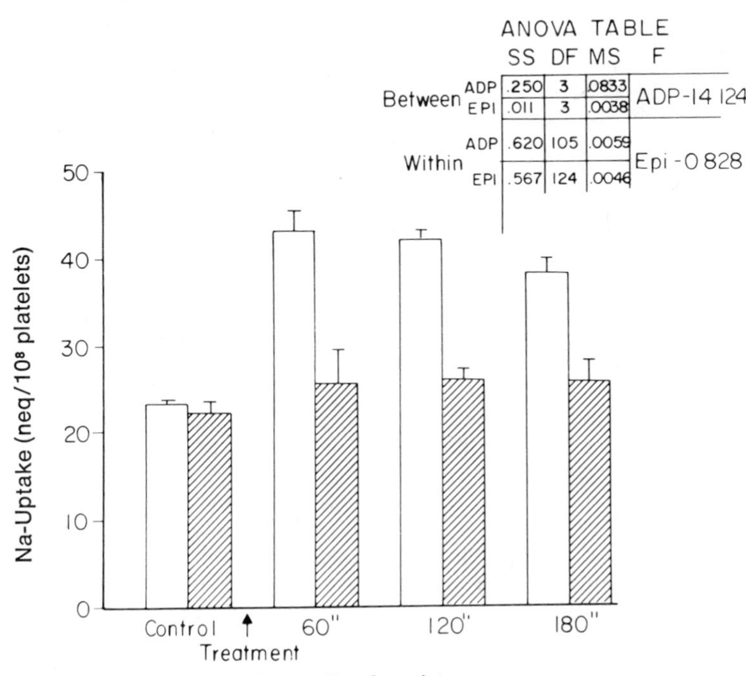

Figure 5. Effect of epinephrine on Na^+ uptake by platelets. Uptake of Na^+ (neq/ 10^8 platelets) is based on an average value of 165 meq/l of Na in citrated plasma. Aggregation initiated by ADP (□, 10 μM) or by epinephrine (▨, 10 μM); values represent mean ± SEM of 10 experiments. (Reproduced, with permission, from Ref. 40. Copyright 1981, Elsevier Biomedical Press B.V.)

rise in platelet Na^+ levels could function to activate platelets by displacing Ca^{2+} from bound sites. This possibility is consistent with the findings of Steiner and Tateishi (42) that Ca^{2+} sequestration by platelet membranes is inhibited by Na^+ and enhanced by K^+. In this connection we found that ouabain, which increases the Na^+ level in platelets, induces an increased sensitivity to both ADP and to epinephrine activation (39).

Involvement of Ca^{2+} in potentiated aggregation. It has been known for some time that low doses of epinephrine will potentiate platelet aggregation induced by ADP (43,44). We next investigated whether these same distinct mechanisms of ADP and epinephrine are operative during potentiation of aggregation by each agonist (45). On this basis we examined whether the mechanism of epinephrine potentiation is due to alterations in membrane permeability to Ca^{2+}. It was found that a requirement for epinephrine potentiation is indeed a transmembrane Ca^{2+} flux, since the potentiated response was completely abolished by verapamil. This inhibition was in turn partially reversed by Ca^{2+} supplementation. Based on these results, it was proposed that low doses of epinephrine induce an uptake of Ca^{2+} which promotes the ability of ADP to redistribute Ca^{2+} from internal Ca^{2+} stores. This potentiation could proceed through a process of calcium-induced-calcium release as has been suggested to occur in the sarcoplasmic reticulum of muscle cells (46).

Support for this notion was provided in a study which investigated the effects of low dose epinephrine on intra-platelet Ca^{2+} binding. Using intact platelets and the Ca^{2+}-sensitive probe CTC, we found that sub-aggregatory doses of epinephrine can indeed lead to a mobilization of intraplatelet Ca^{2+}, stores (47) (Figure 6). An essential feature of this "epinephrine-stimulated Ca^{2+}, redistribution" was a transmembrane Ca^{2+} flux, since the mobilization process was blocked by verapamil. Thus, it appears that an initial increase in intraplatelet Ca^{2+}, as a consequence of Ca^{2+} uptake, can in turn promote the release of Ca^{2+} from internal binding sites. The significance of the internal Ca^{2+} release observed with epinephrine is unclear. Presumably it does not constitute a major source of the increase in cytosolic Ca^{2+}, since D_2O which would be expected to stabilize the DTS does not block epinephrine-induced aggregation. Rather, this ability of Ca^{2+} to stimulate additional Ca^{2+} release may serve as an amplification mechanism by which the effect of Ca^{2+} influx is enhanced.

Ca^{2+} mobilization by metabolites of arachadonic acid. One possible mechanism by which this amplification process could proceed is through Ca^{2+} activation of enzymes involved in arachadonic acid (AA) metabolism. In this regard, it is known that blood platelets metabolize endogenous AA to the potent platelet stimulant thromboxane A_2 (TXA_2) (48). Thus, AA which is esterfied in the phospholipid pool of the platelet is liberated by the action of phospholipase C (or A_2) (49,50). The free AA is then acted upon by a cyclo-oxygenase resulting in the production of the prostaglandin endoperoxide, prostaglandin G_2 (PGG_2). PGG_2 is in turn rapidly converted to prostaglandin H_2 (PGH_2) which serves as substrate for the enzyme thromboxane synthetase. The rate limiting step in TXA_2 production appears to be at the level of the phospholipase. Since phospholipase C or A_2 are Ca^{2+}-activated enzymes (49,50), it is possible that the ability of Ca^{2+} to cause additional Ca^{2+} release may, at least in part, be mediated through Ca^{2+}-stimulated TXA_2 production. The TXA_2 so produced may then act to directly release Ca^{2+} from the platelet DTS (51).

This notion was examined by measuring the effects of the cyclo-oxygenase inhibitor, indomethacin, on epinephrine-stimulated Ca^{2+} redistribution. It was found (47), that inhibition of endogenous AA metabolism did in fact substantially reduce the Ca^{2+} mobilization process (Figure 7). Consequently, it appears that at least a portion of the observed Ca^{2+} release stimulated by epinephrine is mediated through TXA_2 production. Furthermore, direct evidence to support the notion that TXA_2 is capable of causing Ca^{2+} release within the platelet was provided by the finding that addition of the stable TXA_2 "mimetic", U46619, to intact platelets causes a mobilization of internal Ca^{2+} stores (Figure 8).

In summary, it appears that physiological stimulants of platelet function, e.g. ADP, epinephrine and TXA_2 all act through a Ca^{2+} mediated process. Recently, Feinstein (52) demonstrated that the ability of thrombin to cause platelet secretion is also related to the release of intraplatelet membrane bound Ca^{2+}. On this basis, it would seem that intraplatelet cytosolic Ca^{2+} as a second messenger is directly linked to the state of platelet activation.

Inhibition of platelet function by 3'-5' cyclic adenosine monophosphate (cAMP)

One of the first indications that cAMP might be involved in the regulation of platelet function was provided by experiments of Marcus and Zucker (53). In these studies it was shown that the addition of cAMP to platelets resulted in inhibition of platelet aggregation. This notion was supported by subsequent studies of Ardlie et al. (54) who demonstrated that phosphodiesterase inhibitors also blocked ADP-induced platelet aggregation. It now appears that many of the agents, in particular the prostaglandins, which inhibit platelet activation, produce their effect through increases in platelet cAMP levels. Thus, numerous investigators have demonstrated that prostaglandin E_1 (PGE_1) is a potent stimulant of the

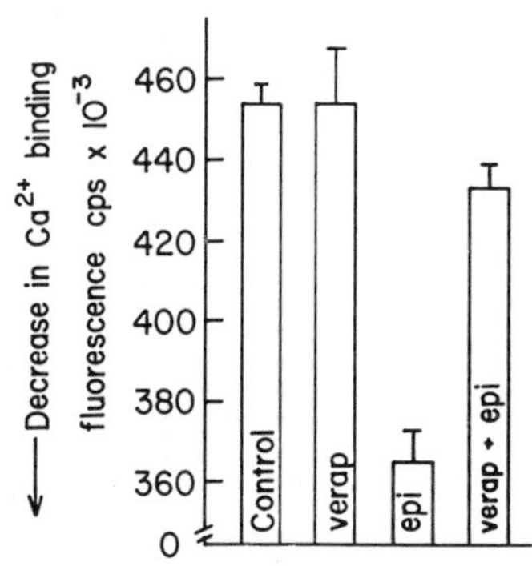

Figure 6. Effect of verapamil on Ca^{2+} mobilization in response to epinephrine.

Human platelets were incubated at 25°C with 50 μM CTC for 40 min. CTC-treated platelets were incubated for 180 s with saline or verapamil (25 μM) alone, saline plus epinephrine (0.1 μM), or verapamil (25 μM) plus epinephrine (0.1 μM). Platelets were then pelleted through silicone oil, and pellet fluorescence was determined as in Figure 1. A decrease in fluorescence counts relative to control indicates mobilization of platelet Ca^{2+}. The fluorescence values are represented as counts/s and are the mean ± SEM of 16 samples from four separate blood donors.

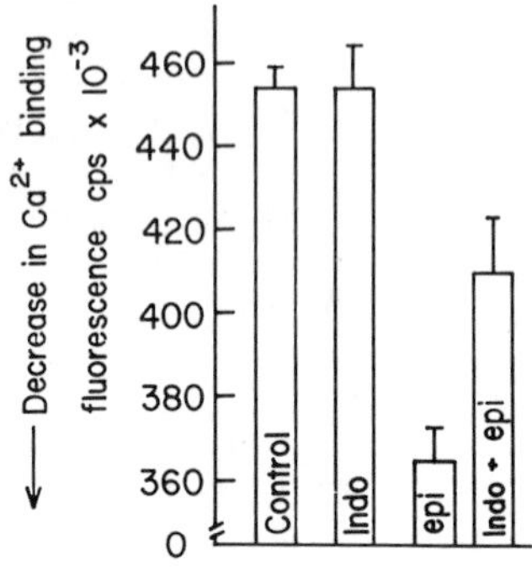

Figure 7. Effect of indomethacin on Ca^{2+} mobilization in response to epinephrine.

Platelets were incubated with CTC and fluorescence determined as described in Figure 6. CTC-treated platelets were incubated with saline or indomethacin (20 μM) alone, saline plus epinephrine (0.1 μM), or indomethacin (20 μM) plus epinephrine (0.1 μM). A decrease in fluorescence counts relative to control indicates mobilization of platelet Ca^{2+}. Fluorescence values are represented as counts/s and are the mean ± SEM of 16 samples from four separate blood donors.

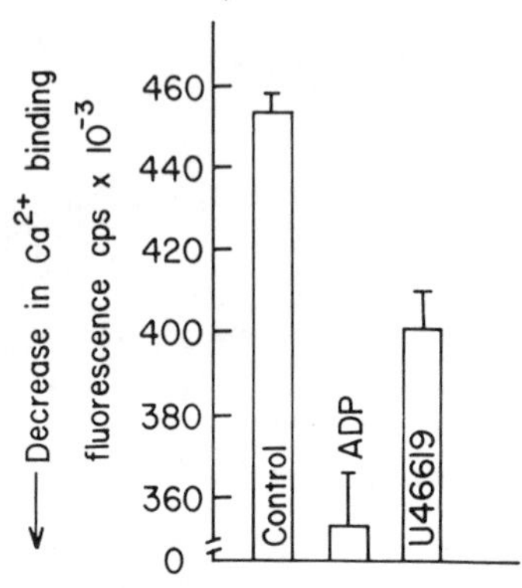

Figure 8. Calcium mobilization in response to U46619.

Platelets were incubated with CTC and fluorescence determined as described in Figure 6. Platelets were then treated with saline (control), ADP (0.5 μM), or U46619 (0.5 μM). A decrease in fluorescence counts relative to control indicates mobilization of platelet Ca^{2+}. The fluorescence change in response to the ADP is indicated for comparative purposes. Fluorescence values are represented as counts/s and are the mean ± SEM of 16 samples from four separate blood donors.

platelet adenylate cyclase (55-59). Furthermore, more recent studies have indicated that the inhibitory properties of prostaglandin D_2 (PGD_2) and prostacyclin (PGI_2) are also primarily due to stimulation of adenylate cyclase activity (60,61,62). Additional evidence to support this concept is provided by the finding, that the effects of the inhibitory prostaglandins are markedly potentiated by phosphodiesterase inhibitors (63). Finally, the influence of cAMP on platelet reactivity is clearly demonstrated by the finding, that an increase of only 25% in platelet cAMP levels is associated with a complete inhibition of maximal aggregation induced by either AA or U46619 (64).

At present then, there is a substantial body of evidence which suggests that increases in platelet cAMP levels are directly linked to inhibition of platelet function. In spite of this, however, the mechanism by which these inhibitory effects are produced remains unclear.

Relationship between Ca^{2+} and cAMP

There have been two theories proposed which attempt to explain the underlying mechanism by which cAMP inhibits blood platelet function. The first model suggests that cAMP is directly involved in the regulation of cytosolic Ca^{2+} levels. Using an isolated platelet membrane fraction rich in the DTS, Kaser-Glanzmann et al. (65) demonstrated that the Ca^{2+} accumulating activity of the isolated vesicles was markedly stimulated by cAMP. These results provided evidence that cAMP may act to promote the movement of Ca^{2+} from the platelet cytosol into the platelet DTS, and were consistent with the previous suggestion of White et al. (13). The net effect of this uptake mechanism would be a decrease in the availability of free Ca^{2+} for the stimulation of shape change, aggregation and secretion.

Direct support for this concept was provided by experiments measuring Ca^{2+} binding in intact platelets. It was found that alterations in platelet cAMP levels, induced by either PGE_1 or PGI_2, were directly related to intraplatelet Ca^{2+} sequestration (47) (Figure 9) as indicated by CTC fluorescence. Furthermore, the effects of PGE_1 or PGI_2 on cAMP levels and Ca^{2+} binding were augmented by pretreatment of the platelets with the phosphodiesterase inhibitor RO 1724. In addition, it was found that the increases in cAMP, stimulated by either PGE_1 or PGI_2, were associated with an inhibition of Ca^{2+} mobilization induced by epinephrine, A23187 or U46619 (Figure 10). Thus, it appears that one consequence of increased cAMP production is a "stabilization" of intraplatelet Ca^{2+} pools. In this connection, however, these experiments did not differentiate between inhibited Ca^{2+} release or enhanced Ca^{2+} resequestration. Consequently, it is possible that cAMP prevents the discharge of Ca^{2+} from the DTS in a manner analogous to the proposed mechanism of inhibition by local anesthetics (66),

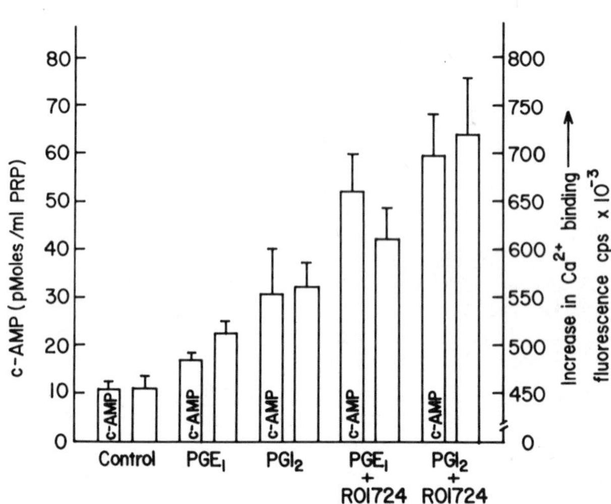

Figure 9. Effect of PGE_1 or PGI_2 on Ca^{2+} binding and platelet cAMP. (Reproduced, with permission, from Ref. 47. Copyright 1981, The American Physiological Society.)

Platelets were incubated with CTC and fluorescence was determined as described in Figure 6. The CTC-treated platelets were supplemented with saline or RO 1724 (100 μM) for 60 s prior to the addition of saline (control), PGE_1 (0.1 μM), or PGI_2 (2.6 nM). Samples of the platelet rich plasma were taken 180 s later, and platelet fluorescence (open bars) and cAMP measured. Values represent mean ± SEM of 16 samples from four separate blood donors.

deuterium oxide (29), or trimethoxybenzoate compounds (67). Alternatively, it is possible that cAMP promotes the rapid resequestration of released Ca^{2+}, such that even in the face of continued Ca^{2+} release, cytosolic Ca^{2+} levels remain low.

Evidence to support the latter mechanism of action is derived from the findings that: 1) cAMP or PGE_1 (which increases cAMP production) stimulates Ca^{2+} uptake into isolated platelet vesicles (65,68). If the sole effect of cAMP were to inhibit Ca^{2+} release upon stimulation, such enhanced uptake would not be observed. 2) Many of the inhibitory effects of cAMP can be overcome by high concentrations of A23187 (68-72). Consequently, if the rate of Ca^{2+} release is greater than the rate of Ca^{2+} resequestration, cytosolic Ca^{2+} levels will rise. 3) In Figure 10 it can be seen that PGE_1 or PGI_2 treatment actually increased Ca^{2+} binding above control levels. If cAMP were only acting to block the release of Ca^{2+} in the stimulated platelet, such increases in Ca^{2+} binding would not be expected. Furthermore, in the presence of PGE_1 or PGI_2 plus a phosphodiesterase inhibitor, epinephrine (which increases platelet membrane permeability to Ca^{2+}) induced a further increase in Ca^{2+} binding (Figure 10). Thus, Ca^{2+} influx induced by epinephrine would provide additional Ca^{2+} for cAMP promoted Ca^{2+} sequestion within the platelet. 4) Low doses of platelet agonists, e.g. ADP, AA, U46619, etc. induce reversible aggregation. Consequently, a mechanism for the resequestration of Ca^{2+} (or Ca^{2+} efflux) must exist in the platelet (13,73).

In this connection, it has been shown that PGI_2 not only inhibits platelet aggregation but also reverses the aggregation process (74). This effect can be readily explained on the basis that cAMP-stimulated Ca^{2+} resequestration exceeds the rate of Ca^{2+} release. Furthermore, it has also been demonstrated that the specific TXA_2/PGH_2 receptor antagonist, 13-azaprostanoic acid (13-APA) (75), causes reversal of both platelet shape change and aggregation in response to AA or U46619 (76,77). Since 13-APA is capable of blocking the release of Ca^{2+} in response to TXA_2 in isolated platelet vesicles (78), the ability of 13-APA to reverse platelet activation is presumably due to a sudden inhibition of TXA_2-stimulated Ca^{2+} release. This inhibition, in the face of continued Ca^{2+} resequestration, ultimately leads to a lowering of intraplatelet Ca^{2+} levels (Figure 11). That the mechanisms of platelet deactivation by 13-APA and PGI_2 are indeed separate, is demonstrated by the finding that the ability of 13-APA to cause deaggregation is potentiated by PGI_2 (79). Consequently, the combined effect of each individual mechanism of inactivation, i.e. inhibited Ca^{2+} release with 13-APA, and cAMP promoted Ca^{2+} resequestration with PGI_2, is more effective in reducing cytosolic Ca^{2+} levels than either mechanism alone.

These observations strongly support the concept that cAMP inhibition of platelet function is mediated through enhanced

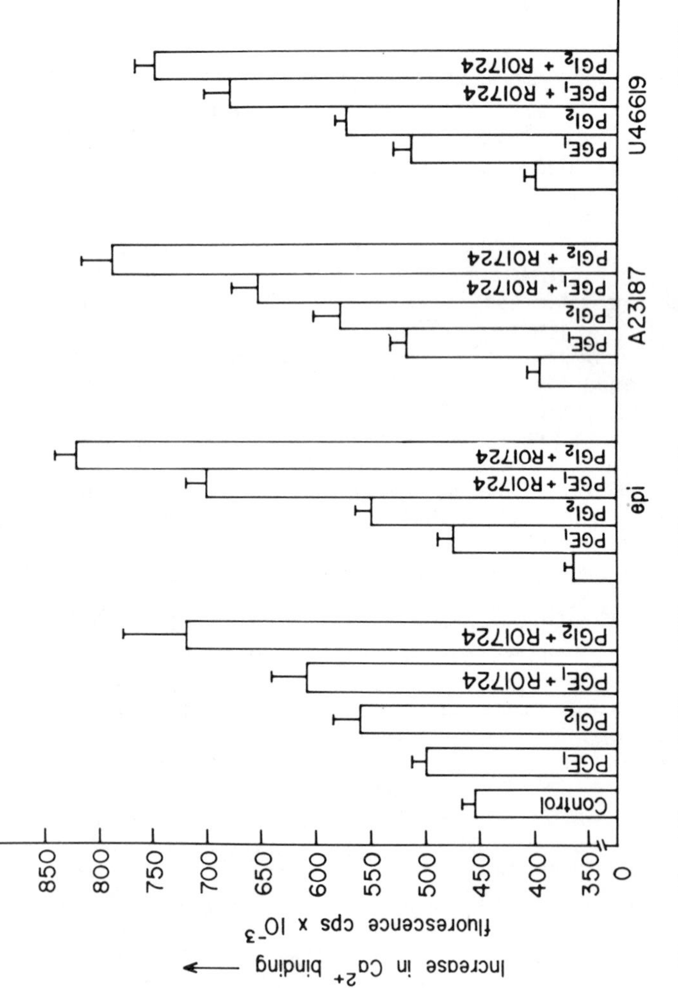

Figure 10. Effect of PGE_1 or PGI_2 on Ca^{2+} mobilization in response to epinephrine, A23187 or U46619. (Reproduced, with permission, from Ref. 47. Copyright 1981, The American Physiological Society.)

Platelets were treated with CTC and fluorescence determined as described in Figure 6. CTC-treated platelets were then supplemented with saline or RO 1724 (100 μM). After 60 s, saline (control), PGE_1 (0.1 μM), or PGI_2 (2.6 nM) was added to the plasma. Platelet fluorescence values are represented as counts/s and are the mean ± SEM of 16 samples from four separate blood donors.

Ca^{2+} sequestration. Although the mechanism by which cAMP may promote the binding of Ca^{2+} within the platelet is unknown, this uptake mechanism has been associated with the phosphorylation of a specific 22,000 dalton protein (72,80). In this regard, it has been postulated (80) that platelets contain an ATP-dependent Ca^{2+} pump, and furthermore, that the efficiency of this pump is enhanced through phosphorylation of a 22,000 molecular weight protein by a cAMP dependent protein kinase.

An alternative mechanism by which cAMP may act to inhibit platelet function was proposed by Hathaway et al. (81). In this model (Figure 12), it is suggested that an increase in cAMP results in inhibition of myosin phosphorylation and consequent inhibition of platelet contractile activity (since unphosphorylated myosin cannot interact with actin). Thus, it was proposed that cAMP causes the activation of a protein kinase which in turn phosphorylates myosin kinase. In the phosphorylated form, myosin kinase is less capable of binding calmodulin and therefore is not as effective in phosphorylating myosin.

Although this model would explain the ability of cAMP to interfere with platelet activation, the results to date were obtained using purified platelet myosin kinase. Consequently, more conclusive proof to support this mechanism of inhibition would require the demonstration that cAMP dependent phosphorylation of myosin kinase occurs under more physiological conditions, e.g. in intact platelets or platelet membrane fragments.

In addition to the possible interactions of Ca^{2+} and cAMP outlined above, certain experimental results have also suggested that Ca^{2+} may actually regulate cAMP production. In this connection, it was demonstrated by Rodan and Feinstein (82) that Ca^{2+} is a potent inhibitor of adenylate cyclase (Ki = 16 μM) in isolated platelet membranes. Support for this concept comes from two lines of evidence: 1) agents which increase cytosolic Ca^{2+} levels e.g. ADP, epinephrine, TXA_2, etc., reduce the elevation in cAMP stimulated by either PGE_1 or PGI_2 (59,83); and 2) the Ca^{2+} antagonist TMB-8 inhibits the ability of TXA_2 to lower cAMP (83). The physiological significance of these findings is unclear. However, previous results have suggested that Ca^{2+} may be involved in other positive feedback systems enhancing platelet reactivity, e.g. Ca^{2+}-induced-Ca^{2+} release or Ca^{2+} activation of phospholipase C or A_2. The reported ability of Ca^{2+} to inhibit adenylate cyclase activity may therefore represent an additional amplification mechanism, whereby slight increases in cytosolic Ca^{2+} levels lead to additional net Ca^{2+} mobilization. In this case, Ca^{2+} inhibition of cAMP production would promote platelet activation by either decreasing the rate of Ca^{2+} resequestration or decreasing phosphorylation of myosin kinase.

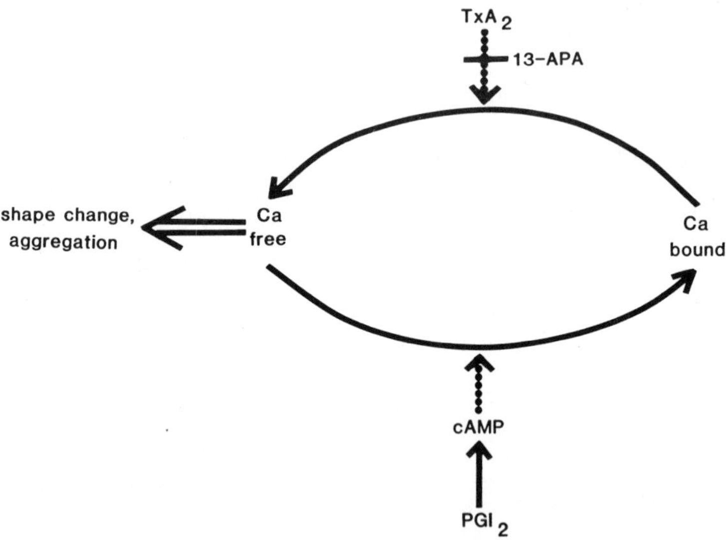

Figure 11. Possible mechanism of calcium release and sequestration.

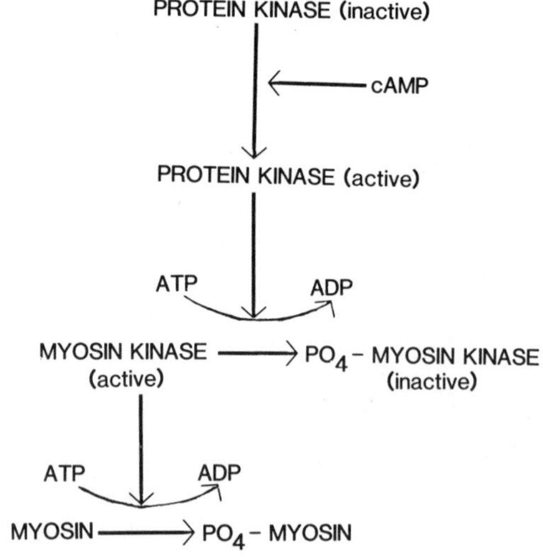

Figure 12. Scheme according to Hathaway et al. (81).

Conclusions

The above findings therefore suggest that there is an intimate relationship between Ca^{2+} and cAMP in the human blood platelet. In this respect, evidence has been presented which indicates that many, if not all, physiological platelet agonists mediate their effects through a common intracellular messenger, i.e. ionic Ca^{2+}. Furthermore, it is clear that the mechanisms for this intraplatelet Ca^{2+} mobilization are indeed complex. Thus, an initial increase in intraplatelet Ca^{2+}, whether as a consequence of enhanced Ca^{2+} membrane permeability induced by epinephrine or direct Ca^{2+} release by U46619, results in additional Ca^{2+} mobilization by activation of AA metabolism or by Ca^{2+}-induced-Ca^{2+} release. The ability of cAMP to modulate platelet function also appears, at least in part, to be due to regulation of cytosolic Ca^{2+} levels, possibly through direct stimulation of Ca^{2+} sequestration.

Acknowledgements

This work was supported in part by a Grant-in-Aid from the American Heart Association. The expert typing services of the Word Processing Center are greatly appreciated.

Literature Cited

1. Holmsen, H.; Day, H.J.; Storm, E. Biochim. Biophys. Acta 1969, 186, 254-266.
2. Moore, S.; Pepper, D.S.; Cash, J.D. Biochim. Biophys. Acta 1975, 379, 370-384.
3. White, J.G. "Platelet Aggregation"; Caen, J., Ed.; Masson et Cie: Paris, 1971; 15-52.
4. Behnke, O. J. Ultrastruct. Res. 1965, 13, 469-477.
5. White, J.G. "The Circulating Platelet"; Johnson, S.A., Ed.; Academic Press: New York, 1971; 45-121.
6. Behnke, O. Anat. Rec. 1967, 158, 121-8.
7. White, J.G.: Am. J. Pathol. 1972, 66, 295-312.
8. Grette, K. Acta Physiol. Scand. 1962, 56 Supplement 195, 1-93.
9. Gordon, J.L.; Milner, A.J. "Platelets in Biology and Pathology"; Dingle, J.T., Ed.; North-Holland: New York, 1976; 3-22.
10. Holmsen, H. "Platelets, Production, Function, Transfusion and Storage"; Baldini, M. and Ebbe, S., Eds.; Grune and Stratton: New York, 1974; 207-220.
11. Massini, P; Luscher, E.F. Biochim. Biophys. Acta 1974, 372, 109-121.
12. Feinman, R.D.; Detwiler, T.C. Nature London 1974, 249, 172-173.

13. White, J.G.; Rao, G.H.R.; Gerrard, J.M. Am. J. Path. 1974, 77, 135-149.
14. Worner, P.; Brossmer, R. Thrombosis Res. 1975, 6, 295-305.
15. Feinstein, M.B.; Fraser, C. J. Gen. Physiol. 1975, 66, 561-581.
16. Murer, E.H.; Stewart, G.J.; Rausch, M.A.; Day, H.J. Thrombos. Diathes. Haemorrh. 1975, 34, 72-82.
17. Kinlough-Rathbone, R.L.; Cahil, A.; Packham, M.A.; Reimers, H.J.; Mustard, J.F. Thrombosis Res. 1975, 7, 435-449.
18. Pressman, B.C. Fed. Proc. 1973, 32, 1698-1703.
19. Reed, P.W.; Lardy, H.A. J. Biol. Chem. 1972, 247, 6970-7.
20. Le Breton, G.C.; Feinberg, H. Pharmacologist 1974, 16, 699.
21. Massini, P.; Luscher, E.F. Biochim. Biophys. Acta 1976, 436, 652-663.
22. Le Breton, G.C.; Dinerstein, R.J.; Roth, L.J.; Feinberg, H. Biochem. Biophys. Res. Comm. 1976, 71, 362-370.
23. Caswell, A.H.; Warren, S. Biochem. Biophys. Res. Comm. 1972, 46, 1757-1763.
24. Hallet, M.; Schneider, A.S.; Carbone, E. J. Memb. Biol. 1972, 10, 31-44.
25. Luchra, M.; Olson, M.S. Biochim. Biophys. Acta 1976, 440, 744-758.
26. Caswell, A.H. J. Memb. Biol. 1972, 7, 345-364.
27. Caswell, A.H.; Hutchinson, J.D. Biochem. Biophys. Res. Comm. 1971, 42, 43-9.
28. Macfarlane, D.E.; Mills, D.C.B. Blood 1975, 46, 309-320.
29. Le Breton, G.C.; Sandler, W.C.; Feinberg, H. Thrombosis Res. 1976, 8, 477-485.
30. Kaminer, B.; Kimura, J. Science 1972, 176, 406-7.
31. Menche, D.; Israel, A.; Karpatkin, S. J. Clin. Invest. 1980, 66, 284-291.
32. Eastwood, A.B.; Grundfest, H.; Brandt, P.W.; Reuben, J.P. J. Membrane Biol. 1975, 24, 249-263.
33. Sandow, A.; Pagala, M.K.D.; Sphicas, E.C. Biochim. Biophys. Acta 1976, 440, 733-743.
34. Owen, N.E.; Feinberg, H.; Le Breton, G.C. Am. J. Physiol. 1980, 239, H483-8.
35. Born, G.V.R.; Feinberg, H. J. Physiol. 1975, 251, 803-816.
36. Moake, J.L.; Ahmed, K.; Sachur, N.R.; Gutfreund, D.E. Biochim. Biophys. Acta 1970, 211, 337-344.
37. Feinberg, H.; Sandler, W.C.; Scorer, M.; Le Breton, G.C.; Grossman, B.; Born, G.V.R. Biochim. Biophys. Acta 1977, 470, 317-324.
38. Feinberg, H.; Michael, H.; Born, G.V.R. J. Lab. Clin. Med. 1974, 84, 926-934.

39. Sandler, W.; Le Breton, G.C.; Feinberg, H. Biochim. Biophys. Acta 1980, 600, 448-455.
40. Feinberg, H.; LeBreton, G.C. "Cell Membrane in Function and Dysfunction of Vascular Tissue"; Godfraind, T. and Meyer, P., Eds.; Elsevier/North Holland: New York, 1981; 129.
41. Wencel, J.; Feinberg, H. Circ. 1979, 58, supplement 2, 124.
42. Steiner, M.; Tateishi, T. Biochim. Biophys. Acta 1974, 367, 232-246.
43. Ardlie, N.G.; Glew, G.; Schwartz, C.J. Nature 1966, 212, 415-7.
44. Born, G.V.R.; Mills, D.C.B.; Roberts, G.C.K. Proc. Physiol. Soc. 1967, 17, 43-4.
45. Owen, N.E.; Le Breton, G.C. Thrombosis Res. 1980, 17, 855-863.
46. Fabiato, A.; Fabiato, F. Nature 1979, 281, 146-8.
47. Owen, N.E.; Le Breton, G.C. Am. J. Physiol. 1981, 241, H613-9.
48. Hamberg, M.; Svensson, J.; Samuelsson, B. Proc. Natl. Acad. Sci. 1975, 72, 2994-8.
49. Rittenhouse-Simmons, S. J. Clin. Invest. 1979, 63, 580-7.
50. Lapetina, E.G.; Cuatrecasas, P. Biochim. Biophys. Acta 1979, 573, 394-402.
51. Gerrard, J.M.; Butler, A.M.; Graff, G.; Stoddard, S.F.; White, J.G. Prostaglandins and Med. 1978, 1, 373-385.
52. Feinstein, M.D. Biochem. Biophys. Res. Comm. 1980, 93, 593-600.
53. Marcus, A.J.; Zucker, M.B. "The Physiology of Blood Platelets"; Grune and Stratton, Inc.: New York, 1965.
54. Ardlie, N.G.; Glew, G.; Schultz, B.G.; Schwartz, C.J. Thrombos. Diathes. Haemorrh. 1967, 18, 670-3.
55. Robison, G.A.; Arnold, A.; Hartmann, R.C. Pharmacol. Res. Comm. 1969, 1, 325-332.
56. Wolfe, S.M.; Shulman, N.R. Biochem. Biophys. Res. Comm. 1969, 35, 265-272.
57. Zieve, P.D.; Greenough, W.B. Biochem. Biophys. Res. Comm. 1969, 35, 462-473.
58. Marquis, N.R.; Vigdahl, R.L.; Tavormina, P.A. Biochem. Biophys. Res. Comm. 1969, 36, 965-972.
59. Salzman, E.W.; Levine, L. J. Clin. Invest. 1971, 50, 131-141.
60. Mills, D.C.B.; Macfarlane, D.E. Thrombosis Res. 1974, 5, 401-412.
61. Gorman, R.R.; Bunting, S.; Miller, O.V. Prostaglandins 1977, 13, 377-388.
62. Best, L.C.; Martin, T.J.; Russel, R.G.G.; Preston, F.E. Nature 1977, 267, 850-2.
63. Mills, D.C.B.; Smith, J.B. Biochem. J. 1971, 121, 185-196.

64. Hung, S.C.; Robin, B.M.; Venton, D.L.; Le Breton, G.C. Circulation 1980, 62, 405.
65. Kaser-Glanzmann, R.; Jakabova, M.; George, J.N.; Luscher, E.F. Biochim. Biophys. Acta 1977, 466, 429-440.
66. Feinstein, M.B.; Fiekers, J.; Fraser, C. J. Pharm. Exp. Ther. 1976, 197, 215-228.
67. Charo, I.F.; Feinman, R.D.; Detwiler, T.C. Biochem. Biophys. Res. Comm. 1976, 72, 1462-7.
68. Fox, J.E.B.; Say, A.K.; Haslam, R.J. Biochem. J. 1979, 184, 651-661.
69. Chaiken, R.; Pagano, D.; Detwiler, T.C. Biochim. Biophys. Acta 1975, 403, 315-325.
70. Feinstein, M.B.; Becker, E.L.; Fraser, C. Prostaglandins 1977, 14, 1075-1093.
71. Rittenhouse-Simmons, S.; Deykin, D. Biochim. Biophys. Acta 1978, 543, 409-422.
72. Haslam, R.J.; Lynham, J.A.; Fox, J.E.B. Biochem. J. 1979, 178, 397-406.
73. Statland, B.E.; Heagan, B.M.; White, J.G. Nature 1969, 223, 521-2.
74. Moncada, S.; Gryglewski, R.J.; Bunting, S.; Vane, J.R. Prostaglandins 1976, 12, 715-737.
75. Le Breton, G.C.; Venton, D.L.; Enke, S.E.; Halushka, P.V. Proc. Natl. Acad. Sci. 1979, 76, 4097-4101.
76. Le Breton, G.C.; Venton, D.L. "Advances in Prostaglandin and Thromboxane Research"; Samuelson, B., Ramwell, P.W. and Paoletti, R., Eds.; Raven Press: New York, 1980, 497-503.
77. Parise, L.V.; Venton, D.L.; Le Breton, G.C. Fed. Proc. 1981, 40, 834.
78. Rybicki, J.P.; Venton, D.L.; Le Breton, G.C. Thrombosis and Hemostasis. 1981, 46, 651.
79. Parise, L.V.; Venton, D.L.; Le Breton, G.C. Thrombosis and Hemostasis 1981, 46. 650.
80. Kaser-Glanzmann, R.; Gerber, E.; Luscher, E.F. Biochim. Biophys. Acta 1979, 558, 344-7.
81. Hathaway, D.R.; Eaton, C.R. Adelstein, R.S. Nature 1981, 291, 252-4.
82. Rodan, G.A.; Feinstein, M.B. Proc. Natl. Acad. Sci. 1976, 72, 1829-1833.
83. Gorman, R.R.; Wierenga, W.; Miller, O.V. Biochim. Biophys. Acta 1979, 572, 95-104.

RECEIVED June 28, 1982.

10

Suppression of Atherogenesis with a Membrane-Active Agent—Nifedipine

PHILIP D. HENRY and KAREN BENTLEY

Washington University School of Medicine, Cardiovascular Division, Department of Medicine, St. Louis, MO 63110

We tested the effects of the dihydropyridine nifedipine on atherogenesis in rabbits fed a 2% cholesterol diet. The drug was given orally, 40 mg/day, and control rabbits received placebo. Nifedipine was well tolerated, and evoked peak reduction in mean arterial pressure of less than 12 mm Hg. Plasma total cholesterol after eight weeks before killing the rabbits was similar in the placebo and nifedipine-treated groups, averaging 1,903 ± 138 (n = 13) and 1,848 ± 121 mg/dl (n = 13 ; mean ± SE; P > 0.8). In placebo-treated rabbits, aortic lesions stainable with Sudan IV covered 40 ± 5% of the intimal surface. The cholesterol and calcium concentrations in aortic tissue were 47 ± 5 mg/g protein and 297 ± 18 μg/g protein. In nifedipine-treated rabbits, values for stainable lesions, aortic cholesterol, and aortic calcium were significantly depressed (P < 0.005), and averaged 17 ± 3%, 29 ± 2 mg/g protein, and 202 ± 14 μg/g protein. Thus, the dihydropyridine suppressed structural and biochemical changes of atherosclerosis without reducing the hypercholesterolemic response to the diet.

Current evidence suggests that calcium plays an important pathogenic role in atherosclerosis (1-7). In cholesterol-fed rabbits, anticalcifying and hypocalcemic agents such as diphosphonates (2-5), thiophene compounds (6), and EDTA (7) have been demonstrated to exert antiatherogenic effects without appreciably altering circulating lipids. This study was performed to determine whether nifedipine, a calcium antagonist, suppresses atherogenesis in cholesterol-fed rabbits. Nifedipine and other calcium antagonists inhibit calcium uptake by smooth muscle cells, but unlike previously used agents, they have no calcium chelating

0097-6156/82/0201-0175$06.00/0

properties (8). Results indicate that intracellular accumulation of calcium, a postulated non-specific mechanism of cell death (9, 10), may be important in mediating hypercholesterolemic vascular injury.

Methods

Fifty-six male white New Zealand rabbits purchased from the same vendor (Boswell's Bunny Farm, Pacific, Mo.) weighing 2.3 to 2.6 kg were housed individually under controlled conditions and randomly assigned to four diet and treatment groups: a) standard pellets and placebo, b) standard pellets and nifedipine, c) 2% cholesterol pellets (Nutritional Biochemicals, Cleveland, Ohio) and placebo, and d) 2% cholesterol pellets and nifedipine. The daily ration of pellets was 100 g for all rabbits, and water was given ad libitum. Nifedipine was force-fed, two 10 mg capsules at 7 a.m. and 7 p.m. (40 mg/day), and untreated rabbits received identical capsules without nifedipine (placebo). In each group, five rabbits were selected randomly for the study of arterial pressure during the first and last week of the diet period. A No. 21 pediatric needle was inserted into the central ear artery and taped to the ear (11). Without restraining the rabbit, the needle was intermittently connected to a Gould P23 Db pressure transducer (Gould, Inc., Instruments Division, Cleveland, Ohio), which was placed at mid-chest level and attached to a Gould amplifier/recorder system. At the beginning and at the end of the diet period, blood samples were collected from the central ear artery of all rabbits into tubes containing Na_2EDTA (1 μg/100 μl). The samples were used for the determination of the microhematocrit, and the separated plasma was analyzed for total cholesterol, triglycerides, phosphate, total protein, and albumin with a Gemini automatic analyzer (Electro-Nucleonics, Inc., Fairfield, N.J.). Plasma calcium was measured by atomic absorption spectrophotometry (see below). The rabbits were killed after eight weeks. The thoracic aorta was removed, cleaned of surrounding tissue, and halved with longitudinal cuts through the anterior and posterior walls. Each longitudinal half-aorta was blotted and weighed. One-half was quickly frozen and stored at -70°C, the other laid flat, intimal side up, on clear plexiglas. The preparation was glued around its edges to the plastic with Permabond 910 (Permabond International Corp., Englewood, N.J.), and stained with Sudan IV (Sigma Chemical Co., St. Louis, Mo.) as described by Kramsch and Chan, 1978; Chan *et al*., 1978; and Holman *et al*., 1958 (4, 6, 12). A plexiglas plate covered with translucent paper was placed on the surface of the preparation. The transparent outlines and stained areas of the vessel were delineated with black ink and areas planimetered with a computerized planimeter (Model 9874A, Hewlett Packard Co., Palo Alto, California). The frozen half-aorta was pulverized at liquid nitrogen temperature, and homogenized in 10 ml of chloroform-methanol (2:1, vol/vol) in a Duall homogenizer (Kontes Co., Vineland, N.J.). Fractions of homogenate were dried and used for measurements of calcium by atomic absorption spectrophotometry on a Perkin-Elmer Model 303 apparatus (Perkin-Elmer Corp., Physical Electronics Div., Eden Prairie, Minn.) as previously described (13), and for estimation of protein by the method of Lowry (14). Small weighed tissue samples from selected arteries were fixed in 10% buffered formalin, embedded in paraffin, and stained with hematoxylin-eosin or Verhoeff-van Gieson stain.

The difference between sequential mean values in the same group was evaluated by the t-test for paired comparisons. The difference between mean values for planimetered lesions was analyzed by Wilcoxon's signed rank test. Values relating lesions to tissue cholesterol were assessed by the same test. Differences between group means for variables occurring in all groups were evaluated by computerized analysis of variance using the General Linear Models procedure (15).

Results

1) Response of the Rabbits to the Diet and Drug Regimens. Two cholesterol-fed rabbits, one receiving placebo, the other nifedipine, died of unknown cause. The diet and drug regimens were well tolerated, and weight gain during the eight week period was similar in the different groups (Table I). High-dosed nifedipine (30 mg/kg day) has been previously shown to be well tolerated in rats (16).

Peak effects on mean arterial pressure and heart rate after each dose of nifedipine were -11 ± 3 mm Hg and 44 ± 18 beats/min during the first week of treatment. These effects were transient, values returning to baseline within two hours or less. Hemodynamic effects of nifedipine did not differ significantly between the dietary groups, nor was there a difference in each group between values during the first and last week of treatment.

2) Blood Chemistry. Results of biochemical measurement in blood are shown in Table I. Nifedipine treatment had no effect on plasma cholesterol levels of rabbits maintained on standard diet or 2% cholesterol diet.

3) Structural Changes in Aorta. In cholesterol-fed rabbits, the percentage of the intimal surface covered by Sudan-positive lesions averaged 40 ± 5% (SE) in placebo-treated rabbits (n = 13), and 17 ± 3% in nifedipine-treated rabbits (n = 13; $P < 0.001$) (Figure 1).

Microscopic evaluation of aortic tissue revealed qualitatively similar lesions in the two cholesterol-fed groups, but a quantitative structural analysis was not performed in this study.

BODY WEIGHT, HEMATOCRIT, AND PLASMA CONSTITUENTS
Table I

	STANDARD DIET		2% CHOLESTEROL DIET	
	PLACEBO (n = 14)	NIFEDIPINE (n = 14)	PLACEBO (n = 13)	NIFEDIPINE (n = 13)
Body weight, grams	3201 ± 113 (2591 ± 68)	3268 ± 121 (2524 ± 52)	3058 ± 166 (2551 ± 63)	3150 ± 158 (2590 ± 54)
Hematocrit, Vol %	40 ± 0.6	41 ± 0.5	35 ± 1.1* (41 ± 0.5)	34 ± 0.9* (42 ± 0.6)
Total Cholesterol mg/dl	45 ± 8	41 ± 6	1903 ± 138* (48 ± 7)	1848 ± 121* (46 ± 6)
Triglycerides, mg/dl	90 ± 23	93 ± 25	129 ± 30	136 ± 25
Total Calcium mg/dl	13.5 ± 0.1	13.5 ± 0.2	13.6 ± 0.2	13.7 ± 0.2
Phosphorus mg/dl	5.9 ± 0.3	6.1 ± 4	5.7 ± 4	6.0 ± 3
Total Protein g/dl	5.9 ± 0.4	5.7 ± 0.4	6.0 ± 0.3	5.4 ± 0.4
Albumin g/dl	4.2 ± 0.3	3.9 ± 0.2	4.1 ± 0.3	3.8 ± 0.2

Values (mean ± SE) refer to measurements obtained at the end of the diet period. Values at the onset of the period, indicated in parentheses, are given only if they differed from the subsequent values ($p < .05$, paired t-test). Simultaneous comparisons between the four groups (analysis of variance, see Methods) revealed significant differences only between cholesterol groups and standard diet groups (values with *; $p < .001$).

4) Biochemical Changes in Aorta. The cholesterol concentrations in aortas from rabbits given standard pellets were 6.1 ± 0.3 mg/g protein for the placebo group, and 6.3 ± 0.4 mg/g protein for the nifedipine group (Figure 2). In cholesterol-fed rabbits, values for the placebo and nifedipine groups differed significantly ($P < 0.001$), averaging 47 ± 5 and 29 ± 2 mg/g protein (Figure 2). Treatment with nifedipine altered the relationship between aortic cholesterol accumulation and formation of Sudan-positive lesions. The ratio relating planimetered lesions (percent) to tissue cholesterol (milligrams per gram protein) averaged 0.85 ± 0.05 and 0.58 ± 0.04% (%• g prot/mg) in untreated and treated rabbits ($P < 0.001$) (Figure 3).

In rabbits given standard diet, calcium concentrations in aortic tissue for the placebo and nifedipine groups were 190 ± 11 and 194 ± 14 μg/g protein. In cholesterol-fed rabbits, aortic calcium for the nifedipine group was significantly lower than that for the placebo group, values averaging 202 ± 14 and 297 ± 18 μg/g protein ($P < 0.005$) (Figure 4).

Discussion

Calcium overload as a pathogenic mechanism of cell injury has been incriminated in various disorders of cardiac and skeletal muscle including catecholamine-induced cardiac necrosis (8, 13), myocardial ischemia (8, 13), myopathies (9, 17), and malignant hyperthermia (18). Of interest is that in the majority of these pathophysiological entities calcium antagonists effectively suppress calcium accumulation and cell necrosis (8, 13, 17, 18). Moreover, myopathic Syrian hamsters given a calcium-deficient diet exhibit fewer lesions in skeletal and cardiac muscle (17). Conversely, facilitation of calcium uptake with ionophores or membrane-active toxins such as lysophospholipids or macrolide antibiotics accelerate necrosis of isolated skeletal muscle (19) and rat hepatocytes in culture (10).

Studies showing protective effects of calcium antagonists in syndromes associated with membrane injury, calcium overload, and cell necrosis may partly explain the antiatherogenic activity of nifedipine. Atherogenesis is accompanied by an accumulation of calcium in arterial walls (1-7), and proliferation of smooth muscle cells, a key process in lesion formation, goes hand in hand with cell necrosis (20, 21). Necrosis of foam cells releasing membrane-active lipid, such as cholesterol, may affect the membranes of neighboring cells and accelerate cellular deterioration and turnover (Figure 5). Decreased necrosis associated with intracellular retention of lipid may explain why lesion formation for a given lipid accumulation was significantly decreased in nifedipine-treated rabbits. We have previously demonstrated that a high cholesterol environment sensitizes isolated arteries to the constrictor effects of calcium, consistent with increased calcium uptake after acquisition of membrane

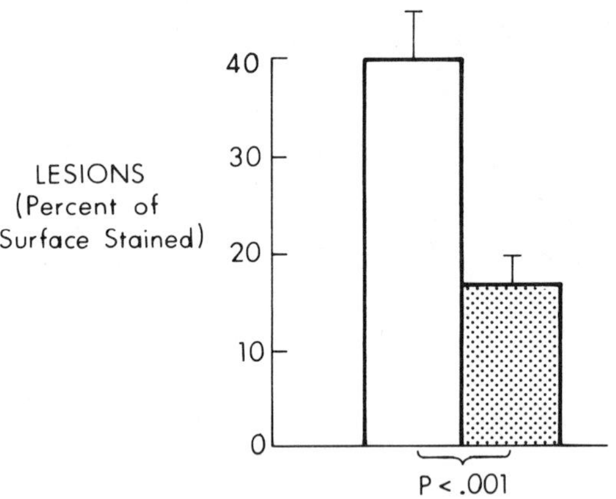

Figure 1. Effect of nifedipine on the extent of aortic lesions (planimetry of sudanophilic lesions) induced by cholesterol feeding. Key (n = 13): □, placebo; and ⛆, nifedipine.

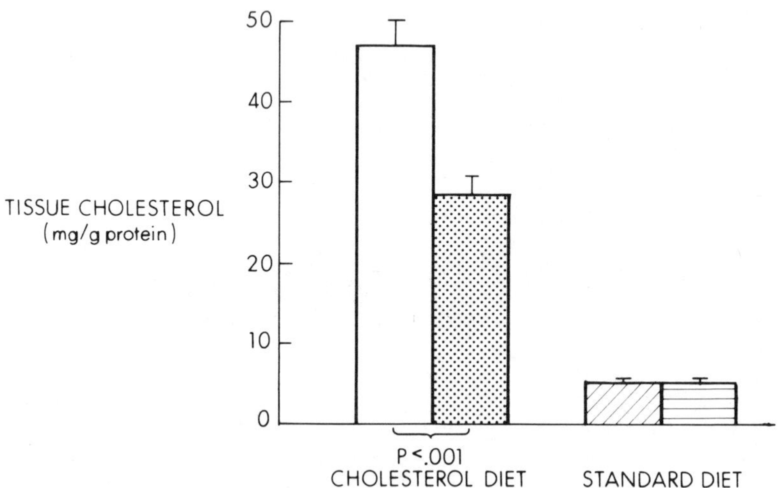

Figure 2. Effect of nifedipine on aortic cholesterol content following cholesterol feeding. Key (n = 13): □, placebo; and ⛆, nifedipine. Key (n = 14): ▨, placebo; and ▤, nifedipine.

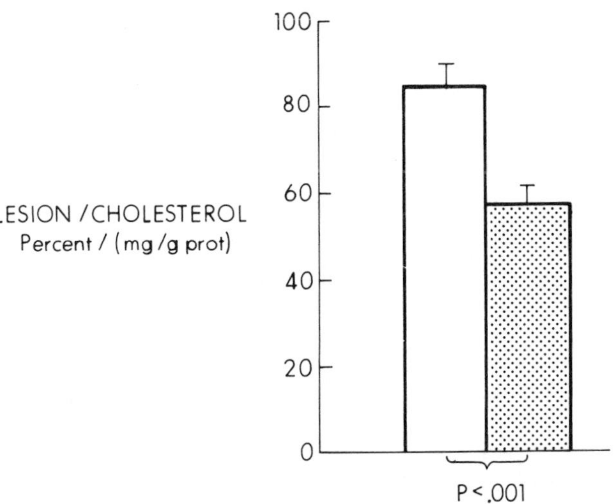

Figure 3. Effect of nifedipine on the relationship between aortic lesions and cholesterol content following cholesterol feeding. Key (n = 13): □, placebo; and ⬚, nifedipine.

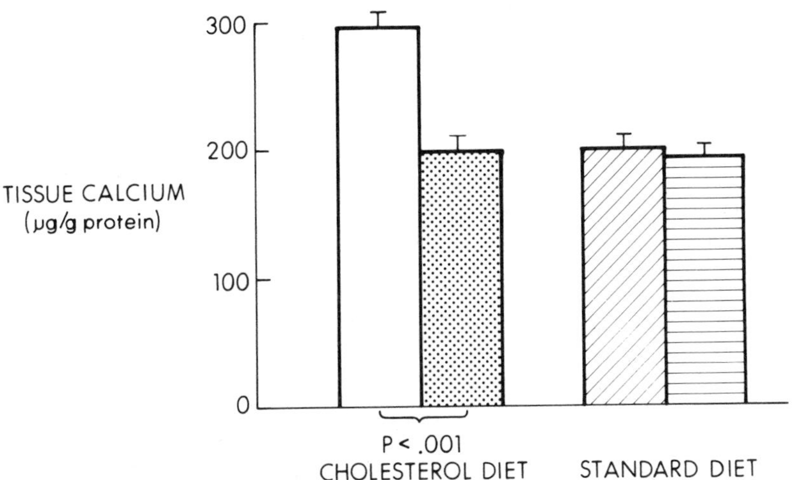

Figure 4. Effect of nifedipine on aortic calcium content following cholesterol feeding. Key (n = 13): □, placebo; and ⬚, nifedipine. Key (n = 14): ▨, placebo; and ▤, nifedipine.

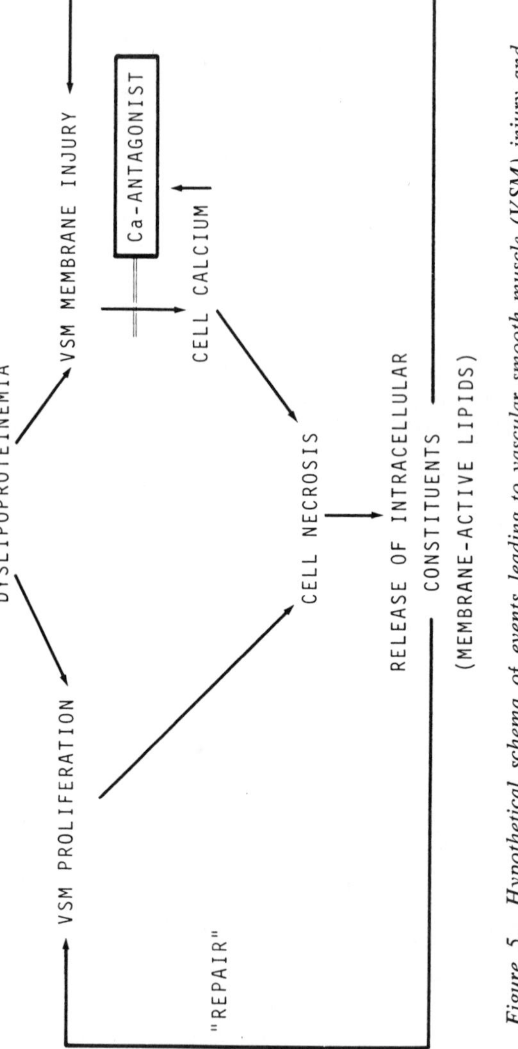

Figure 5. Hypothetical schema of events leading to vascular smooth muscle (VSM) injury and necrosis in the atherosclerotic process.

cholesterol (22, 23). One important question would be to ascertain whether nifedipine inhibits the uptake of lipoproteins by macrophages and smooth muscle cells.

Although nifedipine may influence proliferating smooth muscle, the drug may act by other mechanisms. Arterial pressure is an important determinant of atherogenesis, and elevated pressure aggravates atherosclerosis in cholesterol-fed rabbits (11). Therefore, vasodilator-induced hypotension might exert protective effects, although hypotensive responses were not sustained in this study. Moreover, potent vasodilators such as nifedipine might influence arteries by increasing vasa vasorum flow (24). Calcium antagonists may act on platelets. However, nifedipine does not appear to have antiaggregating effects on human platelets (25, 26), and the action of verapamil and diltiazem on platelets occur only at high concentrations known to exert nonspecific effects (8, 26, 27).

Since calcium antagonists are used extensively for the treatment of coronary artery disease (8), the present study has potential clinical implications. Unlike anticalcifying drugs (2-7), calcium antagonists are not known to affect bone mineralization, and might be evaluated in children with familial hypercholesterolemia.

Literature Cited

1. Blumenthal, H. T.; Lansing, A. I.; Wheeler, P. A. Am. J. Pathol. 1944, 20, 665-79.
2. Rosenblum, I. Y.; Flora, L; Eisenstein, R. Atherosclerosis 1975, 22, 411-24.
3. Potokar, M; Schmidt-Dunker, M. Atherosclerosis 1978, 30, 313-20.
4. Kramsch, D. M.; Chan, C. T. Circ. Res. 1978, 42, 562-71.
5. Kramsch, D. M.; Aspen, A. J.; Rozler, L. J. Science 1981, 213, 1511-2.
6. Chan, C. T.; Well, H; Kramsch, D. M. Circ. Res. 1978, 43, 115-25.
7. Wartman, A.; Lampe, T. L.; McCann, D. S.; Boyle, A. J. J. Atheroscler. Res. 1967, 7, 331-41.
8. Henry, P. D. Am. J. Cardiol. 1980, 46, 1047-58.
9. Wrogemann, K. Lancet 1976, I, 672-4.
10. Schanne, F. A. X.; Kane, A. B.; Young, E. E.; Farber, J. L. Science 1979, 206, 700-2.
11. Bretherton, K. N; Day, A. J.; Skinner, S. L. Atherosclerosis 1977, 27, 79-87.
12. Holman, R. L.; McGill, H. C., Jr.; Strong, J. P.; Geer, J. C. Lab. Invest. 1958, 7, 42-7.
13. Henry, P. D.; Shuchleib, R.; Favis, J.; Weiss, E. S.; Sobel, B. E. Am. J. Cardiol. 1980, 46, 1047-58.
14. Lowry, O. H.; Rosebrough, N. J.; Randall, R. J. J. Biol. Chem. 1951, 193, 265-75.

15. Barr, A. J.; Goodnight, J. H.; Sall, J. P.; Blair, W. H; Chilko, D. M. "Statistical Analysis System"; SAS Institute: Raleigh, N.C., 1979; pp. 237-63.
16. Garthoff, B.; Kazda, S. Eur. J. Pharmacol. 1981, 74, 111-2.
17. Jasmin, G.; Solymoss, B.; Proc. Soc. Exp. Biol. Med. 1975, 149, 193-8.
18. Britt, B. A. Fed. Proc. 1979, 38, 44-8.
19. Publicover, S. J.; Duncan, C. J.; Smith, J. L. J. Neuropathol. Exp. Neurol. 1978, 37, 544-57.
20. Thomas, W. A.; Imai, H.; Florentin, R. A.; Reiner, J. M.; Scott, R. F. Prog. Biochem. Pharmacol. 1977, 14, 234-40.
21. Stary, H. C. Prog. Biochem. Pharmacol. 1977, 14, 241-7.
22. Yokoyama, M.; Henry, P.D. Circ. Res. 1979, 45, 479-86.
23. Henry, P. D.; Witztum, J. L.; Yokoyama, M. Circulation 1978, 58 (Suppl. II), 297 (Abstract).
24. Heistad, D. D.; Armstrong, M. L.; Marcus, M. L. Circ. Res. 1981, 48, 669-75.
25. Vater, W.; Kroneberg, G.; Hoffmeister, F.; Kaller, H., Meng, K.; Oberdorf, A.; Puls, W.; Schlossmann, K.; Stoepel, K. Drug. Res. 1972, 22, 1-14.
26. Margolis, B.; Lucas, C.; Henry, P. D. Circulation 1980, 62 (Suppl. III), 191 (Abstract).
27. Blackmore, P. F.; El-Refai, M.; Exton, J. H. Mol. Pharmacol 15, 598-606.

RECEIVED June 28, 1982.

11

Calcium and the Secretory Process

J. L. BOROWITZ, DAVID E. SEYLER,[1] and CELESTE C. KUTA

Purdue University, School of Pharmacy and Pharmacal Sciences, Department of Pharmacology and Toxicology, West Lafayette, IN 47907

Secretion occurs throughout the body from nerve ends of the brain and periphery and from endocrine and exocrine glands. Thus secretory processes influence many essential body functions and the biochemistry of secretion and its susceptibility to drug action need to be clearly understood. Mechanical events in secretion are well established but biochemical mechanisms are poorly defined. Calcium plays a key role in secretion and when calcium is prevented from entering secretory cells during a stimulus, secretion generally decreases. Agents which block calcium entry into secretory cells probably vary in effectiveness depending on the nature of calcium entry channels in surface membranes. It may be possible to develop calcium antagonists for selective inhibition of calcium entry into cells of a given secretory tissue.

Secretion of various hormones, digestive enzymes, and neurotransmitters throughout the body is carefully regulated. Movements of skeletal muscle, for example, are brought about by discrete release of the neurotransmitter, acetylcholine, from nerve ends. Digestion of food requires secretion of various enzymes from salivary glands and pancreas. Also, neurons of the brain may be thought of as interconnected elongated secretory cells whose discrete secretion of chemical neurotransmitters is the basis for proper mental function.

Nature of the Secretory Material

Material secreted from cells varies in character to include cationic amines, peptides, enzymes of large molecular weight and

[1] Current address: Medical College of Virginia, Department of Pharmacology, Richmond, Va. 23298

0097-6156/82/0201-0185$06.00/0

liposoluble substances like steroids. Generally, these materials are very potent biologically and are packaged inside small spheres called granules (or "vesicles" in nerves) perhaps to prevent effects on the cells of origin, to avoid degradation of secretory material, and to facilitate secretion. Steroids are an exception presumably because they are too lipid soluble to be stored in granules surrounded by lipoid membranes (1). Therefore, steroids are not packaged in granules, but are synthesized as needed.

Secretion Mechanisms

At present, one of the major incompletely understood processes in biology is how cells extrude granule bound material, a process called "exocytosis." During stimulation of secretory cells, granules migrate through the cytoplasm to the plasma membrane, attach themselves and pour out their contents to the cell exterior (2). Electron micrographs show attachment of granules to plasma membrane during secretion (3) and biochemical studies show that only granule contents are extruded and not cytoplasmic enzymes (4) or granule membrane (5). Still, the exact nature of the biochemical events occurring in secretion is largely unknown.

Two systems (microtubules and microfilaments) assist in mechanical events in the secretory process. Movement of granules from deep within the secretory cell toward the inner surface of the plasma membrane involves microtubules. Microtubules are long straight structures which can rapidly elongate ("assemble") and thereby provide an intracellular "taxi service." Subcellular particles rest on microtubules like cars on a railroad track. The drug, colchicine, interferes with microtubule assembly and inhibits secretion in a variety of cells. To illustrate the connection between microtubules and secretion,a correlation was found between inhibition by colchicine of protein secretion from rat lacrimal glands and interference with microtubule assembly (6). Untreated lacrimal glands contained cells with granules clustered near the cell apex, whereas colchicine-treated glands contained granules scattered throughout the cellular cytoplasm. Although what triggers mictotubule assembly is not known, microtubules appear to be necessary for proper movement of secretory granules during secretion.

Secondly, microfilaments provide mechanical assistance at a final stage of secretion. These structures are composed of contractile proteins, the action of which probably aids in granule attachment to surface membrane or in extrusion of granule contents from the cell. A mold metabolite, cytochalasin B, interferes with microfilament action and also inhibits secretion (7). Cytochalasin B prevents attachment of radiolabeled actin to actin already attached to chromaffin granule membrane surfaces and also limits accumulation of myosin on these same structures (8). These muscle proteins probably serve to orient secretory granules in the proper way to facilitate secretion.

Calcium in Secretion

It has been known for 20 years that secretion of granule bound materials generally requires the presence of calcium in the medium (9). If calcium is omitted from the fluid perfusing a secretory tissue, evoked secretion is abolished or at least diminished. However, not all secretory cells handle calcium in the same way and more recent work emphasizes the role of calcium in membrane stabilization, the influence of magnesium ion on calcium mediated events, and the effect of stimulation frequency on calcium metabolism.

Membrane Stabilization. Calcium not only mediates secretion, but it also "stabilizes" cell membranes, making it more difficult for calcium to enter cells. For example, if adrenal medulla is stimulated with 100 μg/ml of acetylcholine (the physiological neurotransmitter) at various calcium concentrations, catecholamine secretion increases as calcium increases up to at least 17.6 mM calcium (9). By contrast, pancreatic insulin secretion in response to glucose, peaks at 5.5 mM calcium and falls off at calcium concentrations on either side of the peak (10). So the relation between calcium concentration in the medium and the extent of secretion varies with the tissue and depends on the degree of membrane stabilization by calcium, a factor which influences the amount of calcium entering secretory cells.

Heavy calcium deposits clearly exist on the surface of insulin secreting cells of the pancreas (11) but no such deposits are seen on glucagon secreting cells of this tissue (11) or on adrenomedullary cell surfaces (12). It is possible that the density of surface calcium deposits determines the degree of membrane stabilization by calcium.

How does calcium act at the molecular level to stabilize cell membranes? Probably it interacts with phospholipids in the membrane to limit ion permeability since phospholipid must be included in artificial membranes in order for calcium to have a stabilizing effect (13, 14). However, there is no firm evidence that calcium stabilizes native biological membranes by interacting with phospholipids, and the structural or functional impact of calcium on various biological membranes is impossible to predict based on lipid composition (15).

Magnesium Concentration. Another factor which may influence secretion is magnesium in the medium. Foldes (16) has recently emphasized the fact that Mg^{2+} (as well as Ca^{2+}) varies in the serum of different species and that many "balanced electrolyte" solutions such as Krebs Solution, do not have the correct amounts of magnesium. Douglas and Rubin (9) showed many years ago that a relatively low concentration of Mg^{2+} (2 mM) prevented adrenal catecholamine secretion caused by addition of Ca^{2+} to a Ca^{2+}-free perfusing medium, although much higher concentrations of Mg^{2+} (10-20 mM) are generally needed to block evoked secretions. Many studies of the role of calcium in secretion from isolated tissues use solutions with incorrect Ca^{2+} and Mg^{2+} concentrations and therefore deviate from physiological conditions.

Stimulation Frequency. Aside from extracellular magnesium and calcium concentrations, at least one other circumstance affects calcium entry into cells during a stimulus. An interesting relationship exists between frequency of nerve stimulation and calcium concentration in the medium which determines the extent of norepinephrine secretion from cat splenic nerve (17). Generally, secretion is increased when either calcium concentration (as previously mentioned) or stimulation frequency (up to 30 Hz) increases. However, the secretory effect of increasing frequency is greater at low calcium concentrations compared to high calcium levels. Thus, in this preparation, calcium concentrations play a role in modulating changes in secretion related to changes in stimulation frequency.

Calcium Pools Used by Different Cells

Considering such differences between secretory cells as size [a mammalian pancreatic insulin secreting cell is 10 μm in diameter compared to a nerve terminal which is 1 μm or less in diameter (18)] and functional requirements (e.g. speed of secretion and frequency of secretion), it is not surprising that differences in calcium handling occur in different cells. Whereas, some secretory tissues need extracellular calcium to mediate secretion, the exocrine pancreas (which is responsible for producing digestive enzymes) uses a mitochondrial calcium store (19) except for sustained responses in which extracellular calcium is also involved (20). Interestingly, salivary gland secretions, which also serve a digestive function may likewise be independent of extracellular calcium during the early phase of secretion [see (21) involving studies of the relation between ^{86}Rb release and extracellular calcium concentration in rat parotid glands].

Calcium Removal from Cytoplasm

If different cells use calcium from different sources to mediate secretion, then it is likely that methods for removal of cytoplasmic calcium also vary. A plasma membrane calcium pump is important in adrenal medulla to extrude mediator calcium (22). In exocrine pancreas, calcium may be returned to intracellular storage pools, as well as extruded from the cell to terminate secretion.

One mechanism for removal of mediator calcium is by packaging of cytoplasmic calcium within secretory granules (23). In this way, calcium is extruded from the cell along with secretory material. Such a mechanism may account in part for the presence of calcium in milk, since calcium is secreted along with milk protein (24), and also may account for the calcium content of saliva which is important in the formation of dental calculus.

Mechanisms of Initiation of Secretion by Calcium

After calcium enters secretory cells, how does it trigger secretion? This question cannot be answered with certainty, but a few pertinent theories have been formulated. First, calcium may activate certain cytoplasmic enzymes such as protein carboxyl methylase, thereby causing methylation of carboxyl groups on granule membrane surfaces (25). This change in granule membrane surfaces may promote the interaction between granule membrane and plasma membrane and initiate the secretory process.

Another theory is that calcium may not directly activate an enzyme to trigger secretion, but rather may first interact with a cytoplasmic protein, calmodulin. The calcium-calmodulin complex may then cause enzyme activation. The enzyme phosphodiesterase is known to be activated by such a calcium-calmodulin complex (26).

Many studies show that divalent cations promote membrane fusion (27, 28, 29) and thereby may initiate attachment of granules to the insides of plasma membranes during secretion. Actually these ideas (i.e. enzyme activation and fusion of lipoid membranes by calcium), are not mutually exclusive since it is possible that calcium initiates more than one intracellular change to trigger the secretory process.

Calcium Channels in Plasma Membranes

In those secretory tissues where extracellular calcium is necessary for secretion, calcium enters by way of plasma membrane channels. Therefore, the nature of membrane channels is obviously very important. Are the channels uniform on a given cell? Do their characteristics vary from tissue to tissue? Many questions remain unanswered, but several studies suggest that a cell may have more than one type of calcium channel. Although not a secretory tissue, smooth muscle has two types of calcium channel: potential sensitive channels and receptor operated channels (30). So, in this tissue [and probably in secretory tissues as well (31)], the nature of the stimulus may determine which channels are opened, the extent of calcium entry and the extent of the response. A high potassium solution, which is commonly used to activate calcium mediated responses, would open potential dependent channels whereas drugs acting on their respective receptors would open a different set of channels, but cause the same overall response.

Hurwitz et al. (32) extended these ideas by giving evidence, again in a nonsecretory tissue, for two potential dependent calcium channels. These authors showed that the calcium channel associated with the phasic contraction of guinea pig ileal smooth muscle was blocked by lanthanum, but the calcium channel mediating the tonic contraction was not. In this system, both these channels were potential dependent. Hence, a variety of calcium channels may exist on cell surfaces.

Calcium Channel Blockers

If calcium channels are so ubiquitous and so important physiologically, then are there any clinical conditions in which calcium channels do not function properly? Actually, the calcium channel blocking agents currently available clinically - verapamil (Isoptin[R]), nifedipine (Procardin[R]), and diltiazem (Cardiem[R]) are very useful for hypertension, angina, or cardiac arrhythmias. In addition, many commonly used drugs, like barbiturates and nitroglycerine, also elicit some calcium channel blocking effects (33). So calcium channel function may not be normal in certain disease states.

Another important substance which blocks calcium channels is hydrogen ion (see 34). It has been known for several years that secretory tissues generally respond poorly to agonists when the pH of the medium is low. A major result of low blood pH is depression of the central nervous system (35). Secretory mechanisms in brain neurons appear to be inhibited by excessive hydrogen ions. One other naturally occurring substance which appears to block calcium channels is ammonium ion (36). Calcium channel blockade by ammonium may be important in liver disease in which tissue ammonia levels are increased and sometimes coma results (37), probably due to altered secretory mechanisms in the brain.

We know that various stimuli will open membrane calcium channels, but what normally closes these channels once they are opened? One suggestion is that calcium ions *per se* close the channels by acting at the inside surface of the plasma membrane (38). So when calcium ion concentration increases sufficiently inside the cell, calcium channels close and no further calcium entry occurs. Interestingly, this mechanism differs from that which closes sodium channels since the latter are inactivated by membrane potential changes. Thus sodium and calcium channels are very different in that calcium, but not sodium, can enter to initiate secretion even if the cell is depolarized.

Specificity of Calcium Channel Blocking Drugs

When calcium channel blockers are used clinically it is assumed that the major effects are limited to the cardiovascular system, and indeed, few side effects have been reported for these agents. However, calcium channels in the cardiovascular system may not be unique and it can be shown experimentally that several secretory systems are also affected by calcium channel blocking drugs. For example, secretion from adrenal medulla (39), and endocrine pancreas (40), and ^{45}Ca Uptake by rat brain fractions (41), are inhibited by calcium channel blockers.

Central Nervous System Effects of Calcium Channel Blockers

This last section presents recent experiments involving effects of calcium channel blocking drugs in whole animals. It is suggested that the following results reflect modification of brain neurotransmitter secretion by blockade of calcium channels.

A simple, yet sensitive, behavioral test for effects on neuronal function is measurement of spontaneous motor activity in mice. This test revealed that verapamil markedly decreased spontaneous motor activity (Figure 1). Despite the decreased movement caused by verapamil, no loss of muscle coordination was noted when the animals were placed on a rotating rod. That these results are related to calcium channel blockade is supported by the fact that diltiazem, a calcium channel blocker different chemically from verapamil, produces the same effects. Functional calcium channels may be necessary for normal motor activity.

Further studies on the central nervous actions of verapamil were based on reports that pain relief by morphine involves calcium (42). Morphine may act by depressing neurotransmitter release from neurons in the pain pathway. Since neurotransmitter release is a Ca^{2+}-mediated event, depression of release by morphine may involve blockade of extracellular calcium influx. If calcium is indeed associated with morphine analgesia then calcium antagonists, by further depressing calcium influx, might modify morphine-induced pain relief. Figure 2 shows enhancement of morphine analgesia in mice by verapamil as indicated by the number of seconds mice treated with morphine and verapamil were able to stand on a copper plate heated to 55°C compared to mice treated with morphine and saline. Note that verapamil alone had no analgesic effect, possibly because it blocks the wrong calcium channels for initiation of analgesia. This experiment illustrates that subtle drug interactions may occur between calcium channel blockers and other agents.

These results raise an interesting question. Are calcium channels in different tissues sufficiently distinct to allow development of calcium channel blocking drugs relatively specific for given tissues other than the cardiovascular system? Such drugs may have novel tranquilizing or smooth muscle relaxant properties useful clinically as well as experimentally.

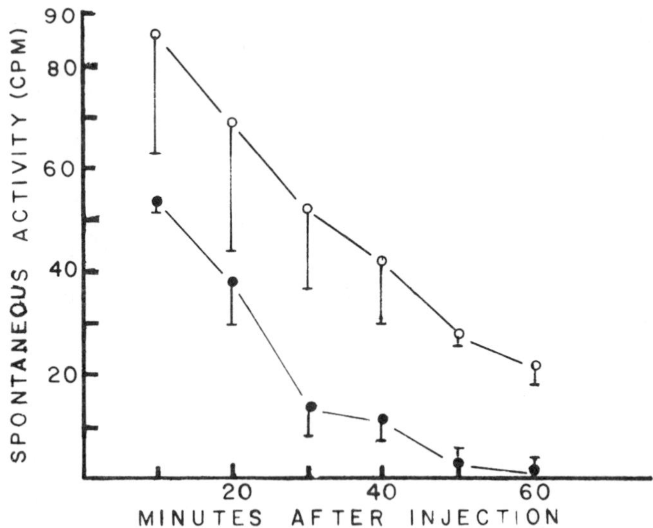

Figure 1. Verapamil-induced decrease in spontaneous motor activity. After a 20-min acclimation period, male mice were given verapamil HCl (●, 10 mg/kg, ip) or saline(○) and placed in activity cages (Woodward Research Corp., Herndon, VA). Twelve mice were tested 10–40 min after injection and six mice were tested 50 and 60 min after injection. The effect is significant at the 1% level.

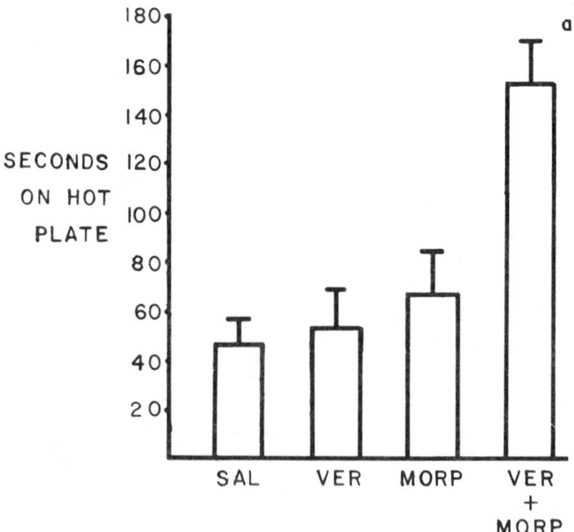

Figure 2. Enhancement of morphine analgesia by verapamil. Saline or verapamil HCl (20 mg/kg, ip) was given to ten male mice followed in 15 min by morphine (2.5 mg/kg, ip). Fifteen min later, the time required for mice to jump from a copper surface heated to 55°C was determined. The effect of verapamil is significant at the 2% level.

Literature Cited

1. Wiechman, B.; Borowitz, J. L. Pharmacol., 1979, 18, 195-201.
2. Wooden, A.; Wieneke, A. Biochem. J., 1964, 90, 498-509.
3. Smith, V.; Smith, D.; Winkler, H.; Ryan, J. Sci., 1973, 179, 79-82.
4. Schneider, F. H.; Smith, A.; Winkler, H. Brit. J. Pharmacol., 1970, 31, 94-104.
5. Trifaro, J. M.; Poisner, A. M.; Douglas, W. W. Biochem. Pharmacol., 1967, 16, 2095-2100.
6. Chambaut-Guerin, A. M.; Muller, P.; Rossignol, B. J. Biol. Chem. 1978, 253, 3870-3876.
7. Douglas, W. W.; Sorimachi, M. Br. J. Pharmacol., 1972, 45, 143-144.
8. Wilkins, J. A.; Lin, S. Biochem. Biophys. Acta, 1981, 642, 55-66.
9. Douglas, W. W.; Rubin, R. P. J. Physiol. Lond, 1961, 159, 40-57.
10. Milner, R.; Hales, C. Diabetologia, 1967, 3, 47-49.
11. Ravazzola, M.; Malaisse-Lagae, F.; Amerdt, M.; Perrelet, A.; Malaisse, W. S.; Orci, L. J. Cell. Sci., 1976, 27, 107-117.
12. Ravazzola, M. Endocrinology, 1976, 98, 950-953.
13. Van Dijck, P.; Ververgaert, P.; Verkleig, A.; Van Dennen, L; deGrier, J. Biochem. Biophys. Acta, 1975, 406, 465-478.
14. Chapman, D.; Urbina, J.; Keough, K. J. Biol. Chem., 1975, 249, 2512-2521.
15. Sklar, L.; Miljanich, G.; Dratz, E. J. Biol. Chem., 1979, 254, 9592-9597.
16. Foldes, F. Life Sci., 1981, 28, 1585-1590.
17. Kirpekar, S.; Garcia, A.; Prat, J. Biochem. Pharmacol., 1980, 29, 3029-3030.
18. Matthews, E. K. "Secretory Mechanisms", Soc. Exp. Biol. Symposium XXXIII, Duncan, C., ed., Cambridge Univ. Press, Cambridge, England, 1979, p. 225-249.
19. Clemente, F.; Meldolesi, J. Br. J. Pharmacol., 1975, 55, 369-379.
20. Williams, J. A. Am. J. Physiol., 1980, 238, G269-G279.
21. Marier, S. H.; Putney, J. W.; Van DeWalle, C. J. Physiol. Lond., 1978, 279, 141-151.
22. Leslie, S. W.; Borowitz, J. L. Biochem. Biophys. Acta, 1975, 394, 227-238.
23. Borowitz, J. L. Biochem. Pharmacol., 1970, 19, 2475-2481.
24. Wooding. F. B. P. Symp. Zool. Soc. Lond., 1977, 41, 1-41.
25. Povilaitis, V.; Gagnon, C.; Heisler, S. Am. J. Physiol., 1981, 240, G199-G205.
26. Cheung, W. Biochem. Biophys. Res. Comm., 1970, 38, 533-538.
27. Igolia, T.; Koshland, D. J. Biol. Chem., 1978, 253, 3821-3829.

28. Paphadjopoulos, D.; Poste, G.; Schaeffer, B.; Vail, W. Biochem. Biophys. Acta, 1974, 352, 10-28.
29. Edwards, W.; Phillips, J; Morris, S. J. Biochem. Biophys. Acta, 1974, 356, 164-173.
30. Bolton, T. Phys. Rev., 1979, 59, 606-718.
31. Bresnahan, S. J.; Baugh, L. E.; Borowitz, J. L. Res. Comm. Chem. Path. Pharmacol., 1980, 28, 229-244.
32. Hurwitz, L.; McGuffee, L.; Little, S.; Blumberg, H. J. Pharmacol. Exp. Ther., 1980, 214, 574-580.
33. Rahwan, R. G.; Piascik, M.; Witiak, D. Can. J. Physiol. Pharmacol., 1979, 57, 444-460.
34. Shanbaky, N.; Borowitz, J. L. J. Pharmacol. Exp. Ther., 1978, 207, 998-1003.
35. Guyton, A. C. Textbook of Medical Physiology, p. 459, W. B. Saunders Co., Philadelphia, 1981.
36. Kuta, C.; Borowitz, J. L. Fed. Proc., 1981, 40, 3592.
37. Schenker, S., Breen, K.; Hoyumpa, A. Gastroenterology, 1974, 66, 121-151.
38. Standen, N. B. Nature, 1981, 293, 158-160.
39. Arqueros, L.; Daniels, A. Life Sci., 1978, 23, 2415-2422.
40. Murakami, K.; Taniguchi, H.; Kobayashi, T.; Seki, M.; Oimomi, M.; Baba, J. Kobe J. Med. Sci., 1979, 25, 237-248.
41. Nachshen, D.; Blaustein, M. P. Mol. Pharmacol., 1979, 16, 579-586.
42. Harris, R. A.; Loh, H.; Way, E. J. Pharmacol. Exp. Ther., 1975, 195, 488-498.

RECEIVED June 1, 1982.

INDEX

INDEX

L

M

N

Q

R

S

T

U

Jacket design by Kathleen Schaner.
Production by Florence H. Edwards and Paula M. Bérard.

Elements typeset by Service Composition Co., Baltimore, MD.
Printed and bound by Maple Press Co., York, PA.